# Bitcoin and Cryptocurrency Technologies

*4 manuscripts*

*Book 1*
*Bitcoin Blueprint*

*Book 2*
*Invest in Digital Gold*

*Book 3*
*Cryptocurrency Trading*

*Book 4*
*Cryptocurrency Investing*

*by*
# Keizer Söze

# Copyright

All rights reserved. No part of this book may be reproduced in any form or by any electronic, print or mechanical means, including information storage and retrieval systems, without permission in writing from the publisher.

**Copyright © 2017 Keizer Söze**

## Disclaimer

This Book is produced with the goal of providing information that is as accurate and reliable as possible. Regardless, purchasing this Book can be seen as consent to the fact that both the publisher and the author of this book are in no way experts on the topics discussed within and that any recommendations or suggestions that are made herein are for entertainment purposes only.

Professionals should be consulted as needed before undertaking any of the action endorsed herein.
Under no circumstances will any legal responsibility or blame be held against the publisher for any reparation, damages, or monetary loss due to the information herein, either directly or indirectly.

This declaration is deemed fair and valid by both the American Bar Association and the Committee of Publishers Association and is legally binding throughout the United States.

The information in the following pages is broadly considered to be a truthful and accurate account of facts and as such any inattention, use or misuse of the information in question by the reader will render any resulting actions solely under their purview.

There are no scenarios in which the publisher or the original author of this work can be in any fashion deemed liable for any hardship or damages that may befall the reader or anyone else after undertaking information described herein.

Additionally, the information in the following pages is intended only for informational purposes and should thus be thought of as universal.

As befitting its nature, it is presented without assurance regarding its prolonged validity or interim quality. Trademarks that are mentioned are done without written consent and can in no way be considered an endorsement from the trademark holder.

# Table of Contents – Book 1

Chapter 1 - Digital Gold…………………………....16

Chapter 2 - Early Investors……………………...24

Chapter 3 - Competition……………………….....30

Chapter 4 - Bitcoin in the born…………………….34

Chapter 5 - The man with the plan…………….39

Chapter 6 - Bitcoin is the FUTURE!....................45

Chapter 7 - Fiat Currency VS Bitcoin…………53

Chapter 8 - Bitcoin VS Gold……………………58

Chapter 9 - Dept Payments ……………………61

Chapter 10 - The power of uniqueness ………..64

Chapter 11 - Bitcoin in a recruit………………...70

Chapter 12 - Mining process of Bitcoin………..75

Chapter 13 - Bitcoin evolution…………………80

Chapter 14 - Purchase Power…………………..88

Chapter 15 - Get ready for the revolution……..93

# Table of Contents – Book 2

Chapter 1 - Why consider Bitcoin……………..119

Chapter 2 – Peer-to-Peer Economy…………...128

Chapter 3 - Is Bitcoin dead?...............................132

Chapter 4 - Will Bitcoin hit $1million?.............136

Chapter 5 – Are you late for Bitcoin?................141

Chapter 6 – 11 Reasons to invest in Bitcoin….145

Chapter 7 – Potential risk of Bitcoin………….151

Chapter 8 – LocalBitcoins…………………….157

Chapter 9 – Hot wallets………………………..161

Chapter 10 – Cold wallets……………………..167

Chapter 11 – Wallet recommendation………..174

Chapter 12 – Bitcoin ATM-s…………………..187

Chapter 13 – Best Online Trading platforms...194

Chapter 14 - Be aware of scammers!..................208

Chapter 15 – Bitcoin Trading…………………221

# Table of Contents – Book 3

**Chapter 1 - Crypto in a nutshell**...................253

**Chapter 2 – Tools to get started**..................262

**Chapter 3 - Bitcoin ATMs**..........................289

**Chapter 4 - Best Crypto Trading platforms**....297

**Chapter 5 – DO NOT GET SCAMMED**……..311

**Chapter 6 – Pump and dump**....................318

**Chapter 7 – ICO-s**..................................326

**Chapter 8 – Identify your investment**………...338

**Chapter 9 – Platform & Applications**………...348

**Chapter 10 – Adjust you portfolio**……………354

**Chapter 11 – Trusted third party VS DIY**…....366

**Chapter 12 – Your portfolio is dropping**……..371

# Table of Contents – Book 4

Chapter 1 – The King of cryptocurrencies........391

Chapter 2 – Litecoin……………………...402

Chapter 3 – Privacy based coins……………...408

Chapter 4 – Monero……………………………412

Chapter 5 – DASH…………………………..417

Chapter 6 – PIVX…………………………........422

Chapter 7 – Zcash…………………………...426

Chapter 8 – Ethereum……………………………431

Chapter 9 – NEO……………………………..447

Chapter 10 – OmiseGo……………………….455

Chapter 11 – TenX……………………………..463

Chapter 12 – CIVIC……………………………472

Chapter 13 – Guide to investing………………477

Bonus Chapter – Filecoin……………………...495

# BITCOIN

## *Blueprint*

## Book 1

### *by*
### Keizer Söze

## Introduction

Congratulations on purchasing this book and thank you for doing so.

This book is an excellent beginner's guide to understand Bitcoin. The contents avoid technical details to provide better understanding to those are new to this technology. There are certain terms that some technical background in Information Technology would help, however, it's not necessary. Everyday english has been used through this book to avoid confusion. This book will take you by the hand to show you, step-by-step, how digital currency was born. For better understanding, first, it begins with a brief introduction of Bitcoin and why often called as Digital Gold. Furthermore, it takes a look at why people are sceptical when it comes to Bitcoin, by analyzing the current market. Next, we take a look at some examples of key individuals, who had an opportunity of getting know Bitcoin at its early days. Followed by explaining the differences between gambling and investing, with the method of understanding some of the key features of Bitcoin technology.
Next, introducing ,,the man with the plan'' aka the inventor of Bitcoin. Next, we comapare Bitcoin to Fiat Currencies, as well precious metals such as

gold, and understand the facts why Bitcoin is so powerful. Once uniqueness of Bitcoin has explained, we exhibit the possibilities of how it can save us from the Dept that inflation has created by old fashioned currencies.

Next, we take a look at how Bitcoin is recruiting the miners, as well what key responsibilities the miners have, and how Bitcoin mining process has been established. Followed by, understand the current market trend, and where Bitcoin already in use, and what are the future plans for further expansion.

In the bonus chapter, we'll go into more detail on how you can get paid in Bitcoin, using Blockchain technology, and how it is extremely beneficial for future of payroll. Finishing by understanding the future of recruitment process, and reputation system that Bitcoin – Blockchain technology can, and already providing us.

There are plenty of books on this subject in the market, thanks again for choosing this one! Every effort was made to ensure the book is riddled with as much useful information as possible. Please enjoy!

## Chapter 1 - Digital Gold

So, you probably heard about Bitcoin, and now thinking to learn more about it. Perhaps, you don't even want to tell your friends or family, only just want to know more about it.

You may be thinking to keep this in secret, and not to talk about it. Some of your friends would laugh at you because you spend your time learning about such thing as Bitcoin.

Some of your relatives would look at you like you are an alien, especially when you mention words like Bitcoin, the blockchain, or cryptocurrency? Some people you know, may tell you things like:,,yeah yeah whatever, I don't care, it's nonsense!,, Some of your friends may act as they care, but, continuously asking questions that have negativity about Bitcoin.

When you ask your friends or relatives and their opinion about investing in Bitcoin, they have no clue what to say, perhaps want to talk you out of it? Do you have a feeling when you talk about Bitcoin to your closed ones, most of them thinking that you are crazy?

In case any of the above mentioned is familiar to you, then don't worry. It's completely normal. The reality is that most people have no clue about the power of Bitcoin, or even what it is. Bitcoin is in the media here and there, mostly, when hackers hit large companies, using techniques like Ransomware, then asking for ransom pain in Bitcoin.

Another famous news, is when Bitcoin is related to terrorists, pornography, drug dealers, arms dealers or some business that's on the dark web, and the only way to pay for those are using Bitcoin. Either way, it is hard to believe that some people never heard of Bitcoin, yet from my experience, it is not exactly correct. Anyhow I have experienced them all, and most responses from people are:

,,I didn't have time to check what is Bitcoin,, or

,,I am too dumb to understand it,, but the best ones are:
,, You should keep this subject to your nerd friends, because I don't care,,

Of course, I have all sort of friends, both investors, as well traders. However, it wasn't always like that, especially when first time I heard of Bitcoin.

Also realized, for some reason people don't talk about it. One of the reasons could be people might hide something, and only want to use this cryptocurrency for bad intentions. I might be wrong. However, the main reason people don't talk about Bitcoin is they don't know what it is, how it works, how to buy, how to spend, how to keep it securely, and so on.

One of my best friends has never heard of Bitcoin or blockchain until I told him. However he is my best friend and all, still just recently said that he doesn't know what will be the outcome. Obviously, he has still no clue what's going on, and that's fine.

My brother has some arguments about Bitcoin, however, when I explained how Blockchain works, which is the platform that allows Bitcoin to run, he has changed his mind, yet still scared of becoming an investor.

My Father has taken keen interests, however, he is very similar to me, and first want to understand how the underlying technology works before he would invest, and this is great!

Another friend of mine is super excited about the idea of investing in Bitcoin. However, he believes the security is number one priority, and he is currently waiting for a Cryptocurrency hardware wallet. He is ordering Ledger Nano.

However, they are completely running out of stock, and the earliest they can deliver is eight weeks ahead. He has decided to learn everything there is to know about Bitcoin in the mean while waiting for the wallet. My opinion on that, he has made the right choice, and it is a wise move.

I will get into wallets, sizes, prices, where to order from, what are the differences, however for now let me introduce Bitcoin, and its glory.

This book is written in August 2017. Therefore if you read it the future, the Market capitalization will be not the same. However right now the Market capitalization of the Bitcoin is $47,679,723,632, and the current price of Bitcoin, against the dollar, is $2891.79

If you ask me on the amount of; 47 Billion dollar +, and still counting? Yes, it's a lot of money that has been invested into Bitcoin alone.

But, why would anyone ever buy Bitcoin? There are multiple answers to that. However, mostly investors, and investing are about building our portfolio.

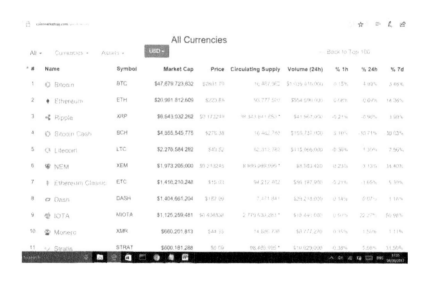

*Note: The screenshot above has been taken on the 4th of August 2017, and the source is www.coinmarketcap.com*

**Why is it a good investment?**

Well, let's look at the quick history of Bitcoin, and its value. I am using the same source from coinmarketcap.com, however, this time I click on Bitcoin to see more details.

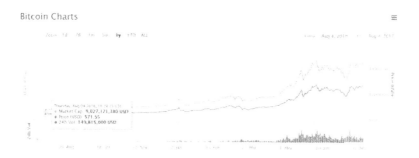

Next, clicking on 1Y on the Zoom field, and now the Bitcoin chart shows me the details of year to date prices. As you can see, exactly a year ago, on 4th of August 2016, the market Capitalization was only $9,027,121,380 billion dollars, however, to round up this figure, let's just say 9 Billion dollars.

If I have to compare this number to the current 47 billion dollars, I could round it up again by saying, only within the last year the market capitalization has became five times more than it was.

In the same time, the value of one Bitcoin was 571.55 dollars. Taking this figure further, I have calculated how much Bitcoin has increased in value over the last year. So again, what I have done is simple; divid 2891.79 by 571.55, and I got an outcome of 5.059. I will round it up to five for better illustration.

However, I believe you can see that the value has increased five times within last year. This means 500% profit. There is no bank I know of, that is capable of doing that, however, if you do know one, please let me know.

Unfortunately coinmarketcap.com is only shows data since 2013, however, another excellent resource in the Crypto world is www.conindesk.com Navigating to coindesk.com, I would like to share with you the value of Bitcoin since it's inception. Unfortunately, even coindesk.com shows only the Bitcoin data since 2010, however, that should be just good enough for the purpose of what I am about to show you.

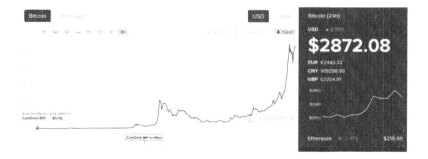

You can see that on the 19th of July 2010, Bitcoin was worth 0.06 dollars. Seven years has past since, and having a quick calculation; you should see that the value has not only five times but 47,868 times has been grown since.

So basically if you would have bought Bitcoin in 2010, for 100 dollars, today it would worth of $4,7 million dollars.

## Chapter 2 – Early Investors

**Roger Ver aka Bitcoin Jesus**

There are many people who have participated purchasing Bitcoin back in the day. They are known as early investors, and today, they are known as Crypto-millionaires. One, that is the most famous worth to mention: called Roger Ver, also known as Bitcoin Jesus.

In 2011, Bitcoin has begun to grow in value, and reached 1 dollar. It was a turning point for many people, and Roger started to learn about Bitcoin and the technology behind called Blockchain. At the same time, he was so convinced that Bitcoin is a Digital Gold, that he has begun to invest. Roger had invested 25,000 dollars into Bitcoin when the price was around 1 dollar. He has deployed some of it, however, even he kept half of it't Bicoin, still that could possibly worth around 35-50 millions of dollars. Roger has multiple companies, as well many start-ups, where he has invested heavily over the years.

All though, some of his involvements are not always favour of everyone in a Bitcoin community, Roger gets into trouble by people misunderstanding his intentions.

Some believe that Satoshi Nakamoto had a different plan, such as the Bitcoin should be completely decentralized, however, Roger also got involved in some Cryprocurrencies where the intentions are to centralize some of those crypto coins.

What people seem to misunderstand, is that Roger is not Jesus, but a business man. Again this is what some people just can't understand, therefore Roger has been called Bitcoin Anticryst too.

Anyhow, Roger is mainly known for a title of Bitcoin Jesus, and I wanted you to know by introducing him, as once you will get involved more in the playground of Bitcoin, his name will keep on coming up from different sources again, and again.

**Winklevoss Brothers**

Yes, they have some involvements to Facebook, even till today, however, they also have a few involvements in Cryptocurrencies, as well Bitcoin.

In 2012, they met a friend who had a friend of another friend who has introduced them the idea of Digital Currency. Cameron has been interested, and began to learn about it. Next he has quickly have presented the idea to his brother Tyler.

Together, they have came to the conclusion that Bitcoin is either a bogus or something big that will change the world. Around 2012, the value of Bitcoin was around 100 dollars when they began to invest slowly. One thing lead to another, they have started to invest heavily, and by the end of the year, they were able to buy up 1% of all Bitcoin in circulation. At the time they have invested around $11 million dollars buying Bitcoin, however, that value has grown at least 25-30 times by today. In US dollar today 2017 they have as much Bitcoin that worth between $275 million to $300 million dollars.

**Steve's story**

There are just too many people who have become wealthy. However, there are also speculations around Bitcoin. For example one of my friends, Steve, just recently told me his story what happened back in 2013.

Steve has never believed in Bitcoin until it reached the value of $100. But then suddenly Bitcoin went crazy, and the value has changed to $200.

So, he said he was going to wait for the right moment to invest, and it meant for him that once Bitcoin goes back to $100, he will invest for $10,000 dollars worth. He has kept on waiting, and waiting then suddenly the value of Bitcoin has changed from $200 to 300 dollars.

He didn't know what to do, if he should wait or not, so he has invested into Bitcoin, buying 10 Bitcoin for

3000 dollars in November 2013. However in the same month, Bitcoin has reached $900, and he has sold all his Bitcoin. Steve has made a good profit, however, he told me recently that he was very lucky.

I told him that those Bitcoin would worth lots more now, and if he would have kept on hold of it, he could have a better portfolio, however he couldn't care less.

**Gambling vs. investing**

The reality, is there are many early investors, and there are plenty of success stories around these people, yet some may call them early gamblers.

The fact is that no one knows what the future brings. However, people do their best, and those not afraid, will succeed. I have learned it for a while now, and it is easier than you think.

To overcome fear, one of the best solution, is to be confident. There is no point to be confident in a decision that you have not done any research on, therefore it is always vital to understand what exactly you invest into. Again, it's a straightforward process. Do your research, learn as much as you can until you have no doubt. In order to reach that point, you will know what you have to do.

Believe me, it can go in two ways, either you are confident what you know through the process of continuous learning, or you are confident that the particular project doesn't make you confident enough to invest, and move on to another project.

# Chapter 3 – Competition

The reality is, that in 2015 to 2016 Bitcoin had less value, and pretty much was moving on the scale of between $300 - 600 dollars. However, in those two years, there were many other cryptocurrencies born, in fact, hundreds of other cryptocurrencies arrived to the Crypto market to compete with Bitcoin.

So far no luck for any other cryptocurrencies to beat Bitcoin, and by navigating back to coinmarketcap.com, as of today, there are 1032 other currencies exist. As you can see I have included in the screenshot the date; 4th of August 2017. Those who think that one or two other cryptocurrencies can take over Bitcoin, I think it's pretty evident that for nine years, thousand + currencies just couldn't succeed.

Not only couldn't do it, but Bitcoin became even stronger in value, in fact, the highest of them all. These are simple facts, and even bad reputation by criminal usage, or hundreds of fake news, just can not change. Bitcoin does not care what's in the news, or what people opinion is; it looks after itself, makes corrections, and so far it seems, the sky is the limit.

The beauty of making a profit on Bitcoin, is there is no tax to pay. There is no company called Bitcoin, and there is no server called Bitcoin, neither Bank called Bitcoin.

Bitcoin is running on a completely decentralized peer-to-peer network, where, there is no master node or primary server, therefore there is no boss. It's running on a technology called blockchain, where, in every 10 minutes, there is a new block get's validated by proof of work, using mathematical computation power, using very sophisticated Cryptography known as Elliptic Curve.

Bitcoin's algorithm also includes the discrete logarithm problem, that not only provides one of the best security in the world, but creating an un-hackable system.

**Note: More detailed guide on Block creation, can be found in the Blockchain books, particularly technical guide in Volume 2 – Mastering Blockchain.**

What you have to understand is that Bitcoin is here to stay, and not going anywhere. It is unstoppable, therefore people, governments, banks or financial institutions, like it or not.

You might have heard the sentence sounds like this: ,, If you can't beat them, you have to join them.,,

Well, that is what exactly banks do now. Not only investing heavily to Cryptocurrencies like Bitcoin or Ethereum, but creating their digital currencies.

One of the most famous called: Ripple. Ripple is already on the market with the current capitalization of $6.6 Billion being on the third place.

| # | Name | Symbol | Market Cap | Price | Circulating Supply | Volume (24h) | % 1h | % 24h | % 7d |
|---|---|---|---|---|---|---|---|---|---|
| 1 | Bitcoin | BTC | $47,082,505,533 | | | | | | |
| 2 | Ethereum | ETH | $20,926,889,818 | | | | | | |
| 3 | Ripple | XRP | $6,683,960,257 | | | | | | |

You have to ask the question: Are they scared of Bitcoin? Or they just see that it's an excellent opportunity for them too?

It is indeed a wonderful opportunity for everyone. No matter who you are, an individual, a bank, a criminal, it works for everyone.

# Chapter 4 – Bitcoin in the born

So, why Bitcoin? There are many other cryptocurrencies on the market, however, for some reason, Bitcoin seems to be the winner of all. The answer might be simpler than you think. It is the most wildly implemented Cryptocurrency, and the reality is that most people look at cryptocurrency, as Bitcoin only. I have some many friends who never heard of any other cryptocurrency except for Bitcoin. People just never heard of other currencies, however most of them did hear about Bitcoin.

Furthermore, I already touched on the dark web, and Bitcoin is what everyone accepts on there, however it the other hand, accepting Bitcoin has been legalized in many countries recently. In Japan, it's been announced, that by the end of Summer 2017, there will be 260,000 Stores that will accept Bitcoin as a paying method.

Along with Japan, there is another major country Russia, who also looking at legalizing Bitcoin as a legal paying method. I love the idea of choice: cash, card, or crypto, when you about to pay in the supermarket. As there are more and more countries becoming interested in the idea of Bitcoin Payment, the rest of the world will catch up too.

Before we get into more detailed facts, let's understand how the idea has developed over the years, until we reached to implement Bitcoin.

In the 1970's there was a lot of interests in Cryptography, especially in how mathematical algorithms can change the world, by using it as money, or at least, as part of a payment method. Diffie Hellman Key Exchange was very promising, in fact, we still use the same methods for our current online banking for security.

**Note: Diffie Hallmen Key exchange explained in my previous Book: Mastering Blockchain - Volume 2**

Unfortunately, when it comes to digital data transfer, a new problem has been introduced, that is known as double spending. Just think of it like I have a digital photo that I am emailing you. What happens is we now both having the same image, unless I delete it. However, you would never find out if I ever did delete it. Then you could easily send the same image to some of your friends or family members; therefore we could easily duplicate digital data, and we still are. However double spending problem has been resolved over the years, and now it has it's own platform. Running as an application in a software form, on the top of the internet, and the technology called Blockchain.

**Double spending problem**
Double spending problem is the main issue that needed to be solved to introduce a new electronic money system. The problem could have been solved by using a central trusted third party online, that could verify the electronic cash has not been spent yet.

Back in the day, the idea was that this trusted third party could have been anyone like a bank, broker, or any entity, someone who can facilitate interactions between two sides. Anyone, however the problem was that trust the third party, would be still required. Of course, there are plenty of disadvantages for trusting in third parties, in fact in any financial services.

In 2008, when financial crisis hit, several banks failed, taught us, there is no such thing as trusted third party. They have failed mainly because of mismanagement, greed, or even because of involvement of illegal bank activities. Furthermore, half of the adults around the world have no access to financial services, because financial institutions are too far away or too expensive to use. Third parties are commercial entities; therefore, they will charge fees for their services.

If you think about inventing a new electronic money, one of your goals should be to make it accessible to anyone around the world. Third parties have the power to suspend customers accounts. For example, a few years ago PayPal has suspended WikiLeaks donation account, and froze its assets.

PayPal claimed, WikiLeaks to encourage others to engage in illegal activity. This was not a result of legal process but rather, the result of fear of falling out of the favor with Washington. Third parties can also deny or limit access to your assets. For example in 2015, the in Greece, the banks have limited access for cash withdrawal because of the rush on the banks.

**Double spending solution**
The solution for double spending without third party now exist, and that is what blockchain allowed for Bitcoin. Bitcoin was the first application which has solved the double spending problem without

the use of third parties, or having any involvement with any centralized system. Satoshi Nakamoto has came up with an idea of Bitcoin, and created its original reference implementation.

Satoshi has solved the double spending problem, using a technology called Blockchain. The system is based on cryptographic proof, instead of trust. Blockchain technology was originally used as a cryptocurrency for the payment transaction between two parties, but nowadays it can be used many other services such as:

- Notary Services,
- Identity services,
- Voting services, and so on

# Chapter 5 – The man with the plan

*Note: I have explained in great details regards to Satoshi Nakamoto and multiple theories as well facts around the real inventor in my book called: Blockchain for beginners Volume 1.*

*I have dedicated four chapters around Satoshi Nakamoto, in Volume starting from Chapter 5 to Chapter 8, however if you have not that book yet, I will provide a little overview of who Satoshi Nakamoto is. In case you have read my book Blockchain for beginners, you may skip to the next chapter now.*

## Satoshi Nakamoto

First, I would like you to understand that this book has been written in the second quarter of 2017. Therefore, by the time you are reading this book, it's possible that new light might be shed on who Satoshi Nakamoto is.

With the current knowledge at hand, let's try to understand who Satoshi Nakamoto is.

First of all, Satoshi Nakamoto is an inventor of Bitcoin, as well the blockchain technology. All through it's a false name, this is how he introduced himself to the internet. It is a men's name.

However, it is possible the Satoshi Nakamoto might be a woman. This is one of the biggest mysteries in the technology world. Yet, most people don't want to know exactly who Satoshi is; nevertheless, they are thankful for the technology he created.

Unfortunately, many people think that because Satoshi Nakamoto has invented Bitcoin and the blockchain technology, he is also the owner of those too. The reality is that Satoshi Nakamoto has no control over the Blockchain—neither bitcoin; therefore, it really doesn't matter who Satoshi Nakamoto is.

But yeah, we still want to know who is behind the curtains; so, let's think about it again. Satoshi Nakamoto is reasonably a man or a woman—of course—he could be a couple, a group of people, or even a group of women for all we know. Satoshi Nakamoto might be ten people together, but also could be a massive team of 100 individuals.

Satoshi Nakamoto might be a child, or he could be old men. Satoshi Nakamoto might have died right after he released his white paper; therefore, he had no time to show his real face.

I do understand if you are getting bored of these accusations, so let's begin thinking in a different perspective. Satoshi Nakamoto might not even be human. Well, you might think of me being over the limit.

However, it's just so odd that we couldn't figure out who Satoshi Nakamoto is in the past decade; not where he resided, but who he is—honestly—we have no idea. Someone might know exactly who he is. However, there is no confirmation that would ever have enough evidence to prove who Satoshi is.

I've always loved to watch sci-fi movies, and I came across one called Arrival. Some of these super old movies, still hold up today. For example, back in the day, some sci-fi stories featured individual objects, or tools that we might use in the future and some that we've already been using for years. I don't want to get into too many specifics; however, think about face time talk back in the 80's. It was a concept that one day we might be able to do that.

And nowadays Skype and Facebook Video Chat is in our daily lives. In fact, there are millions of people connected and capable of being on skype video chat for hours, using our cell phones. The first iPhone was created and launched to the market ten years ago, in 2007. Since we have gone through some dramatic changes, and the next decade will be even more impressive.

So back to the picture called Arrival, I hope that you have seen it already too, and that I will not spoil it for you. However, if you haven't seen it yet, you might want to skip the next few lines.

In the movie, we have received a visit from Aliens that are here to help us by providing visibility in the future. Again, sorry if you have not watched the movie yet, you will probably hate me for this. The concepts of the film are excellent, no wonder it received an Oscar, even though it might have deserved more than that, but that's just my opinion.

When I think about this sci-fi movie, I am thinking about the fact that it is very similar to the same concepts.

We have received a technology called Blockchain from an unknown person—or I should say from an anonymous source—that will change our world dramatically! I wonder how the film creators came across the idea…

I am not suggesting that there are Aliens out there, but I can't deny it either. What I can tell you is that IT Professionals, Software Developers, Experience Programmers, even Cybersecurity Experts are fascinated by this technology, and often refer to it as an ALIAN TECHNOLOGY."

The blockchain is huge, and it certainly takes months, if not years, to fully understand it's technical details, and how it fits together.

Another thing is that, more and more often, it is said that this technology is just too complex for one man to build. Therefore, there is no way that Satoshi Nakamoto was working on it alone.

So back to the million-dollar question, "Who is Satoshi Nakamoto?"

Let's look at some of the claims over the years so that you can decide for yourself.

What you have to understand is that Satoshi Nakamoto went silent in 2009, and remained like that for the next five years, or at least on the forum where he previously posted and was always active.

Note: For further reading on Satoshi Nakamoto, please visit the book called Blockchain for Beginners – Volume 1.

## Chapter 6 – Bitcoin is the FUTURE

Many friends often ask me:
Keizer! – How can you define in simple terms what Bitcoin is?

Well, the answer always depends on who asked the question. If you are completely new to Bitcoin, like I was at first, I even thought at some point that Bitcoin has a physical form. However, Bitcoin is completely virtualized, therefore untouchable. Now that you have a bit of understanding who Satoshi Nakamoto is, you must know that he has created a software called Blockchain; however, the intention of the software was to introduce an application. To run the application, it was required to be run on a network, that is known as a peer-to-peer network.

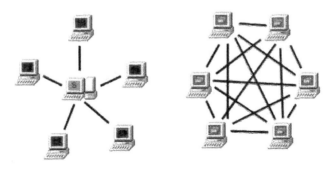

The peer-to-peer network, therefore, has created a system, and it was able to execute the application on the software.

That application's primary purpose is to introduce a new paying method, known as digital currency. Because the application is running on a peer-to-peer network, this is also known as a decentralized network or decentralized system, or I should say the best description for it: decentralized currency.

Because each transaction is required validation through proof or work, it needs to be recorded in the ledger. Therefore it's also known as a decentralized ledger system.

A currency that has no centralized system, it can not be tempered, therefore providing a single source of truth. A single source of truth that is validated by proof of work, therefore it's a technology.

In fact, Bitcoin is running on multiple algorithms, and protocols using the internet, therefore it's a combined into a very sophisticated technology.

Because it's using combined technology, it can't be manipulated. The peer-to-peer network requires internet access, and because it's running on the web, I could also say: it is the currency of the internet. Because nobody owns it, and there is no Bitcoin boss, I could also say that is the money of the people. Sure, the banks can invest too.

However it's not owned by them, and they can not lend it, well at least not yet. Bitcoin account does not require to provide your name, address, race, age, occupation, all there is necessary to have internet access. It is capable of providing anonymity, so I could also call it: Anonymous payment system.

So back to our question: what is Bitcoin? Well, Bitcoin is representing our future, in fact, the future of money, as the transactions will not require banks or any trusted third party, therefore it's not only excellent, but the best solution to do business in the future. Let's look at Bitcoin in a bit more detail.

**Bitcoin is a software**

Well, Bitocin is certainly not hardware. Bitcoin network can be downloaded, and it can run on any computer, that has a connection to the internet. However, it is software.

## Bitcoin is an Application

As I mentioned Bitcoin is software; however, it's running on the technology called Blockchain.

Blockchain already has thousands of other cryptocurrencies running; therefore Bitcoin itself is just one of those many applications. It is fair to mention that Bitcoin was the first application on the blockchain, and many other applications are only trying to copy or compete with Bitcoin.

## Bitcoin is a network

As I mentioned before, Bitcoin is running on a peer-to-peer network. Any computer in the world can participate in the network, as there is no master node or master server. The validation process is running on every node that is part of the system once a new block gets created and added to the Blockchain.

## Bitcoin is a system

Because it's running on the peer-peer network, and using multiple known technologies, it's also known as a system.

## Bitcoin is money

I think you will agree with this right away. Bitcoin is the most wildly implemented and accepted cryptocurrency in the world.

## Bitcoin is a decentralized Cryptocurrency

There are many more decentralized cryptocurrencies, however, again, it is fair to mention that Bitcoin was the first of them all.

## Bitcoin is a decentralized ledger system.

Each transaction is required to have a validation process, as well making sure that each validation is secured in the great ledger; therefore it's also known as a decentralized ledger system.

## Bitcoin is a very complex technology

Bitcoin is running Elliptic Curve Cryptography to participate in a discrete logarithm problem, also using hashing as well ASCII encoding, and many more technologies combined. The reality is that 90% of Bitcoin investors have no clue how Bitcoin works, however they know that is just working fine. If you learn how the underlying technology works, called Blockchain, you will understand why it will change the world.

Many people believe to know what Bitcoin is, however when it comes to the block validation process, as well who receives what amount of transaction fees, and on what algorithm is determining these methods, you would be surprised that very few people understand how Bitcoin really works.

***Note: Further learning on Blockchain attributes and how Bitcoin works in depth, please visit my other book called: Mastering Blockchain – Advanced Guide: Volume 2***

So back to the main question: What Bitcoin is in simple terms:

This is not only my opinion; however I am very confident to say that Bitcoin is the FUTURE!

It will change our lives, in fact already shaping it, and my advice to you is that faster you learn about it, sooner you understand what Bitcoin is capable, easier it will become to deal with it.

I am not only talking about investing, but learning it. There is no point to invest into anything that you do not fully understand, or at least have a high level of understanding where you put your money in the first place.

## Chapter 7 – Fiat Currency VS Bitcoin

Bitcoin is so powerful, yet many people speculate around Bitcoin as well other cryptocurrencies that they are just a bubble.

The reality is that those individuals who say Bitcoin is a bubble, are mistaken. Please don't get me wrong by saying that, as I do know many people who believe that Bitcoin is a bubble, but it is only a reflection of their knowledge of Blockchain.

For example my ex-boss, Andrew, who has high educations as well studying every day, well respected, not only by me, but amongst other Infrastructure, and Application Manager. Still he has not considered Blockchain enough yet, therefore he believes that Bitcoin is a bubble.

When I have explained some of the core functions, of the underlying technology called Blockchain, he has exhibited some frustrating expressions. However, after few days he was asking me about some other functions of Bitcoin, and who is behind the maintenance of the software.

Obviously, I can tell, he has changed his mind of Bitcoin since. In fact, he is consistently reminding me each time when the value of Bitcoin is going up. Before you think Bitcoin is some kind Day trading or any other easily manipulated stock market feature, I warn you now that Bitcoin has nothing to do with the Stock market.

The Stock market opens every Monday to Friday 9 am to 5 pm. Over the weekend, each nights, even weekdays the stock exchange is closed. Opposite Bitcoin is open 24/7, day and night, weekdays and weekends, there is no Bank holidays or day offs. Bitcoin as well the cryptocurrencies are running in a market that always open, and never stops. About every 10 minutes, there is a new block gets validated, and it doesn't matter what time of the day is or which country you or I reside in, new Bitcoin get's mined.

## Who allowed the Bitcoin to hit the market?

This is one of those questions that many people involved and trying to figure out, but the answer is straightforward. There was nobody. There was not a single individual or company who has allowed Bitcoin to exist, in fact, there was no company been asked in the first place. Bitcoin did not ask for permission, and neither was waiting for authorization.

When the European Union together have decided to bring this new currency called: EURO, to the market, was taking years. Meetings, and long discussions on how each EURO should look like and how it should present itself in a physical form, like 5 Euro, 10 Euro, 20 Euro, and so on.

Then, a long meeting of fights and discussions, they had to decide which country is qualified to start using Euro. Next, agreements on possible printing locations, as well once Each Euro will be converted, how much it should worth according to the German Marks, or to the Italian Lira. Each country that was involved, had no choice but to participate in these long meetings of discussions around the currency, and of course many long fights too. Either way, the actual implementation has only happened in some countries according to the original plans. The point is, by the time every little detail has been discussed, the Euro already began to lose its value.

Many people have become poor after the change of currencies, and those, that had all their life time savings kept in a bank, suddenly become worthless. Of course, there was a way to making sure there is a profitable way of getting into the new currency.

However these solutions were kept for the wealthy, and those were able to make a move, which involved the right timing. Euro is a FIAT currency that was manipulated from the beginning of its existence.

It's centralized by the biggest players in Europe. In the European governments, politicians made sure that they will become rich by centralizing the Euro, however in the case of Bitcoin, there was no discussion, neither was any permission asked from those in Governments.

Bitcoin didn't ask for permission, didn't talk to the banks, neither to any politicians, in fact, most of them don't even understand how Bitcoin works. As always, when people don't know something, they don't like it, they hate it, they afraid of it, and slowly, once they understand it, they will respect it, and they will eventually invest in it.

**Why I choose Bitcoin?**

The reality is that Bitcoin might be spelled out in the news as fake money or bubble, however, the reality is that Bitcoin can be expanded dramatically. What I mean is Bitcoin can be easily divided into multiple fractions, while paper money is only causing issues.

For example, imagine that you have to pay for something that only costs 3 dollars in the shop, but you only have a $20 note and the shop owner does not have a change.

When it comes to Bitcoin, it is not a problem, as you can pay as much as you want, and the software will tell you exactly how much Bitcoin is 3 dollars, at the time of the payment.

## Chapter 8 – Bitcoin VS Gold

Again, when it comes to payments, using gold, or any other precious metal, there is a little problem.

There are multiple challenges, so let's think about the storage first. Let's say that I would like to buy a new sofa in the shop, and I know that the price of the couch is 1kg of gold exactly.

Should I take that 1kg of gold into the trunk of my car? Oh no, wait, I have to buy a plane ticket for myself, wife and the kids for our annual summer holiday too.

Those tickets, + Hotel room prices will cost about 0.57 Kg of Gold, so I have to take some gold with myself too right? There is no point in getting into this any further, as Gold or any precious metals simply can not be the solution to the future of money. They are too heavy, difficult to cut them, and the value of products are just a pain to be measured.

It would not work, and it was one of many issues why it was stopped as a paying method in the first place.

**Content type?**

When you think about Music, they used to be in cassette format, then CD, eventually mp3, however for a long time now, we can stream music.

Video contents, are the same, as well books. Back in the day, you had to have DVD player along with DVD-s, as well the physical books had to be carried, in order to access the contents.

However, nowadays you can stream them all. So, why should I take cash everywhere with myself if we can stream it now, like in a case of Bitcoin.

Also, if I travel to another country, I have to change my dollars to euros, or British pounds, instead we could just have Bitcoin everywhere.

Bitcoin does not require bank, neither bank card, all I need is a smart phone.

A smart phone that I am able to stream music, video, books, as well money.

This is an excellent idea. Streaming money is the future like everything else has become streamable, money can to that too.

Technology changes, and we are able to develop things that was none existing a decade ago using technology.

2 decades ago, the idea of streaming any music, or any video using a smart phone was ridiculous, and of course unbelievable.

Today, having a smartphone, all these functions can be done with no issue. Again, streaming money in the future, is not unbelievable, as we can do this already since 2008.

As a final sentence to this topic, streaming money should not be surprising to anyone, as there is nothing new to it, however commercial television and fake newspapers will try to keep this away from us, as long as possible.

Just think about like an average bank would say that they are not good enough anymore to look after your money.

Banks are always get hacked, they always take ages to get the transfers done, they are charging high fees, and they are using your money to pay their debt.

By saying such things, they should just close down right away. However, they will do their best for keep up the current system as long as possibly can.

## Chapter 9 – Debt Payments

### Who is paying the Debt?

Unfortunately, people still believe that only nations can issue currency. However, it's not true.

The reality is that Bitcoin has not asked for permission from any Government, neither from any bank or politician, yet exist.

Many shops already accepting Bitcoin as a paying method, however let's move on to the topic of debt.

In 2017 January, the global debt has hit the new record of $217 Trillion dollars, however since it's continuously growing. It's not only impossible to pay this amount, but this amount will not going to be paid.

So what will happen? Well, according to this figure, it seems that the sky is the limit.

So if you think that this debt can be paid back in precious metal like Gold, it is another misconception.

All the world's Gold that the Earth has, is worth all together $8.5 Trillion dollars.

That's a lot of Gold if you ask me, but let's think again, and calculate how much does all that Gold worth.

You don't have to be good in math to understand that $8.5 Trillion against $217 Trillion does not stand a chance.

So basically, if we all just donate all our Gold for the purpose of debt repayment, it would cut it.

Gold just not good enough to take over the worlds of money, in fact, can not even stand a chance.

Gold has other issues too, like storage as well transporting it from one place to another, therefore as a paying method not even worth mentioning.

Gold for Online payments, that idea, of course, is ridiculous. Lastly, in case we want to create small units for payments, Gold should not even be considered.

## Chapter 10 – The power of uniqueness

I have already explained why FIAT currencies are worthless, and I am sure that you have understood it by now. In fact, everyone knows it, yet the governments are always trying to create another currency, by saying this time it will be different, but they know it will be not. Still, people have no choice, except to go with the flow and use the nation issued currency.

The good news is that we don't have to go with the flow anymore. In fact not only don't need banks anymore, but we can be our own bank. That's all nice, however, the reason that Bitcoin can take over the world of currencies, theatrically very simple, yet it's a complicated process. Bitcoin is unique, not only because it's a cryptocurrency, but Bitcoin has a limited supply. That sounds scary for some, however, let me explain a little further so you can understand it better.

Bitcoin has a limited supply of 21 million. When you think about any other FIAT currency, no one exactly knows how much money has been printed, yet when it comes to Bitcoin, there is an exact figure.

As everything else that is unique and has a limited supply, it's worth more than those products or money that has an infinite supply.

I remember when the Italians had Lira, and the notes were ranging from 1000 to 500,000.

It was ridiculous, as it was nearly worthless in the end when Euro was about to take over.

Just when the Euro has been taken over the Italian Lira, 1000 Lira was equivalent to 0.5 Euros, and the largest notes of 500,000 Lira was worth of 250 Euros. Imagine that you had 1 million Italian lira on you; however, that was only worth of 500 Euros, which is around 590 dollars.

It was silly really. Anyhow, the point is that more Lira was on the market less it was worth, and in the end, it would just become worthless. On the other hand, Bitcoin is very different.

Not only that we know exactly how much Bitcoin will be ever in the market, but we also know by when it will happen. What I mean is not all Bitcoin is currently on the market yet. Instead of flooding the market with a new currency, Bitcoin has been introduced to the market very slowly. It is depending on what year you are reading this book, however currently as of August 2017, there is only 16,499,762 Bitcoin in circulation:

I have just visited coinmarketcap.com to see exactly, however in about every 10 minutes, this figure changes. In case you want to know exactly how much Bitcoin is in circulation at the time you read this book, just visit the following link:

https://coinmarketcap.com/currencies/bitcoin/

## 50 Bitcoin in every 10 minutes

The exact date of the software was in 2009 3rd of January when the first 50 Bitcoin was born. Then for the next for years in every ten minutes, there was another 50 Bitcoin mined.

## 25 Bitcoin in every 10 minutes

From 2012 this has continued; however, the algorithm changed, and from 2012, in every ten minutes there was only 25 Bitcoin mined until 2016.

## 12.5 Bitcoin in every 10 minutes

Since 2016, in every 10 minutes, there is 12.5 Bitcoin is mined until 2020.

## 6.25 Bitcoin in every 10 minutes

From 2020, the Bitcoin algorithm will change again, and for the next four years until 2024, in every 10 minutes, there will be 6.25 Bitcoin mined.

Basically, in about every four years, the Bitcoin algorithm will change, and the amount of Bitcoin that will be mined is kept on getting halved in every four years. This process will carry on until the year of 2140. By then, there will be 21 Million Bitcoin mined, and that will be the last of Bitcoin will be ever mined.

The process of creating Bitcoin is based on Gold mining. What happens with the Gold is something very similar what is the process with the Bitcoin, however Bitcoin is based on digitally, using multiple technologies. Another similarity that Bitcoin has, is to compare it to Gold mining difficulty. There was a time when people have found lot's of Gold mines, and able to mine plenty of them, however over time, mining Gold, became more and more difficult.

As I explained previously, mining Bitcoin is getting twice as difficult in every four years; thus there will be less and less amount to be mined. When you think about this process, you might realize that it also means for each of these four year passes, the value of Bitcoin will be in a continuous increase. I am not able to predict the future, however, obviously, we all can analyze the last eight years.

Bitcoin is based on a technique of Gold mining, however, by now you must be very curious what do I mean by mining. Of course, mining also introduces another question, who the miners are?

## Chapter 11 – Bitcoin in a recruit

This section will introduce who are the miners and their purpose, starting by what was the reason in the first place to create such role as a miner.

Then will move on what exactly their responsibility, and why they have a huge influence on the network.

Ending by how they participate within the mining process, and what miners must do at all times.

*NOTE: This has been explained in other book called: Blockchain for beginners – Volume 1, in Chapter 10, and Chapter 11.*

*In case you are familiar the process of Bitcoin mining and the miner's responsibility, you may go ahead and skip the rest of this Chapter.*

*However, if you want to refresh your knowledge or unaware of the Bitcoin mining process you may carry on reading this chapter.*

## Background

Let's first think about how new value enters the system. Back in 2008, Satoshi Nakamoto only created 50,000 Bitcoin to start the process.

If you think about it, that he built all 21 million in the first place, the bitcoin would be worthless, and the idea would have been dumb. Instead, Satoshi started with a moderate amount of Bitcoin creation.

## The solution

As the Bitcoin community grows, more and more value would be required for the system to be kept alive.

There is a particular process that is needed for the system to be maintained; Satoshi has come up with the solution by creating a role. This solution is not only solving one but two issues:

1. Permanently validating transactions
2. Adding new value into the existing system

## The role is called: Miner.

Miners can be individuals, or any bitcoin citizen. However, over time, many large companies have been formed, such as Genesis Mining, where you, as an individual, can join and rent their mining facilities.

There are many other miners who over the years have created a pool, and many of them also offer to join these pools for certain reasons that I will discuss shortly.

## The responsibility of a miner

First, let me explain why they are called miners and what it is they do. They are called miners as the analogy has been used with gold or any other precious metal.

They work together to create new value, similar to gold miners who are digging underground. However, bitcoin miners are sealing each transaction into the ledger. Therefore, we could call miners, finalizers or authenticators.

To get rewarded for such work, the miners receive bitcoins, and this is how new value is added to the system.

The miners validate, authenticate, certify, and finalize the transactions by specific processes. Once the miners have created a new block that is accepted by the citizens, the record of the transaction cannot be modified, making it permanent information. This will also become irreversible. Therefore, no one can ever challenge it or change it, in the future.

The miners are sealing the blocks, which in itself can take an enormous amount of computing power, assuring that they cannot be easily replicated. There are multiple methods that each miner may use for the validating processes.

Some of the miners may use different software, even creating their own in-house made software to speed up the authentication process. However, it doesn't matter what software they use, as all of their work will be checked.

It starts when a miner begins to gather transactions that have been broadcasted on the network, then starts checking those transactions, and eventually sealing those collections of transfers and operations into a new block.

A miner receives bitcoins as a reward for each sealed block that is added to the blockchain.

# Chapter 12 – Mining process of Bitcoin

**Block creation**

Explaining each block creation can be done in multiple ways; however, some sound very confusing, but it also depends on how much you understand technology. Therefore, hearing or reading it the first time can be difficult to comprehend.

I already explained that miners have an unusual role for validating each transaction in the form of a block. Now, let's discuss what it takes to create each block.

1.
Start a new block. Even if the miners are halfway done validating a block, eventually, they will drop everything and concentrate on starting a new block.

2.
Select a new transaction. This is when the miners are choosing from thousands of operations that are broadcasted over the network.

3.
Check priority of the transaction. This time the miners can go back to number one by starting a new block if they find that the transaction they have selected previously is not that significant. However, if the priority is high, the miners may go on and move to the next step.

4.
Check that the transaction is valid. This is a process that every miner must check, there is no exception of avoiding this step for any miner. However, if the transaction is found to be faked, or not valid, the miners have to stop the process, and go back to number 1 and start a new block and get another, hopefully, valid transaction.

5.
Accept the transaction. If the previous transaction was tested as a valid transaction, it must be accepted.

6.
Seal the transaction. Again, if the transaction has been found valid and accepted, now it's time to seal that transaction.

7.
Add the transaction to the transaction tree inside the block. This process can only be done once all previous steps have been verified.

8.
Check for the size of transactions. The miners need to check if there are enough transactions within the transaction tree, to seal the block. If there are not enough transactions yet, the miner will not be able to seal the block until there are enough transactions. Therefore, the miners must go back to number 2 of selecting a new transaction again, and again, until there are sufficient transactions for sealing the block.

9.
Check interruptions. This is the process where the miner must make sure that no other miners have

sealed the block in the meantime with the same transactions inside the block.

10.
Seal the block. Once there are enough transactions for sealing the block, the miners will seal the block.

11.
Broadcast the block. The miners must broadcast the new block that has been sealed; however, if the miners have been interrupted within the block sealing process, they might have to start a new block all over again.

12.
Start a new block. This is the next step in the process; however, as you see, we are now back to step number 1.

As I mentioned, miners might get interrupted while they are sealing the block and once they broadcast it, if another block has already been sealed by another miner with the same transactions within a block, the block will not be accepted. Therefore, you must start a new block.

Each block is created about every 10 minutes. As a result, 144 blocks are created each day. As I mentioned before, the miners who have successfully added a new block into a blockchain get rewarded a degree of bitcoin.

The reward for each new block creation used to be 50 bitcoins from 2009 until 2012. The reward for a new block gets halved every four years; therefore, from 2012, until 2016, the award for each new block used to be 25 bitcoins.

Currently, since 2016, until 2020, the reward to a miner for a new block that is added to the blockchain is 12.5 bitcoins; however, from 2020, it will be only 6.25 bitcoins until 2024. This process will be continued until 2140 until the last bitcoin will be created.

## Chapter 13 – Bitcoin evolution

The question around what Bitcoin can become is indeed what many people concerns, as well excited about. Bitcoin can stop many people's current job, and especially those are working in Banks or financial institutions.

Banks are not happy about Bitcoin, as there is a constant fear that Bitcoin can steal their job anytime shortly.

Bankers, as well stock Brokers are in fear, however, most of them getting the feeling of safety as they have begun to invest in Bitcoin, as well learn about, what Bitcoin is becoming shortly.

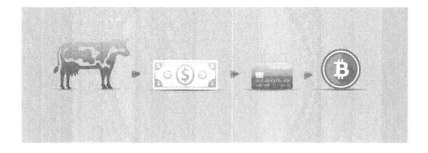

Its' ok to be the sceptic; however there is a limit and people should realize this by looking back to the history of recent technology that created evolution, as well a revolution of today. Yes, I am talking about the internet; however, there are other notable technologies too that I will mention in this chapter.

**Internet**

It stands for interconnected networks, of course, you are probably known that already. It is known for other names too such as www – aka world wide web, net or web.

At first, it was slow and painful, there is no doubt about that. However it has evolved, and there were more and more websites to visit. Still, there was a problem, you must have typed the website address into your browser to visit a particular page.

As I said it was painful, and people were sceptical that this is not going to work. Imagine if you have mistyped a website address, you got nowhere, then you had to figure out what did you do wrong.

So it wasn't making any sense, it was more like a pain to find any website, and for many people the internet wasn't just it. There was a belief that the internet will not work ever. As always, when there is a problem, some of us keep thinking about a solution. The solution was Google.com, a search engine where you can type in few words that you are interested about, and you will get related pages.

That was the first part of the evolution of the internet; however more people have begun to use the internet, more it was believed that it would not survive all that traffic.

Again, a solution was introduced, and the internet has started to scale, where internet service providers have born. Those ISP-s – Internet Service Providers have been providing additional internet routes for end users for faster connection and reliability.

Once emails were introduced and began to scale by having an increasing number of users, it was another new belief, that the internet would not be able to handle that traffic, and the e-mail system will kill the internet.

Of course, it was stabilized very quickly, however people began to attach files, and images to the e-mails, and this new type of communication seemed to be a killer of the web.

Just imagine the traffic growth on those networks where few email attachments have made the actual email size by 10, even 20 times bigger. Emails used to take a day to get it delivered; however, this issue has been overcome again, by extending the network with better and newer technology.

Once voice and video over IP ( Internet Protocol) were introduced, people got so scared that they believed the web would not only can't be scaled, but it will shut down.

This was also a false prediction, as we know the world today, or at least in 2017 when this book was written. We have multiple services where we can pay a subscription, or even for free, able to stream voice and videos anytime and the quality is excellent.

One of the most famous video conferences that are freely used is Skype; however, there are much more like it, such as WhatsApp or Facebook Messenger. Once it comes to video content like Vimeo or youtube, you know that these channels just work all the time and there is no issue with the quality that the 'current ISP-s are providing.

The internet as we know today is where every large company has moved their businesses, including banks and large shops. You can buy holidays, book Hotels, shop pretty much anything anytime. As you can see the internet has changed dramatically over the last two decades, and it has completely changed our lives and the way we think.

**Linux**

Yes, it is an operating system that apparently not many can handle, yet let me discover you, how Linux has changed our life. Linux is a free open source software, that was created in 1991 by Linus Torvalds; however, it is based on a Unix operating system.

Unix dated back to the end of 1960-s, however it has become known only in the 90-s to the world. Linux has so many different kinds of flavours that are just endless, in fact, you can create your version of Linux due to its open source code.

However, back in the 90-s, it was known as a worthless free software, as anyone can download a copy for free and make modifications to it, had no real value.

When people hear Linux as an operating system often get scared and run away. It is a misconception and a wrong belief that Linux OS is a complicated system to handle. 2 decades later, nearly everyone has at least one Linux based device. The reality is that back in the day, Linux was the favourite operating system of all computer nerds, including me.

It had no desktop version at first; therefore anything you wanted to do, was based on the CLI – Command Line Interface.

Yes, that's right, a black screen with white letters on it. No desktop, no start menu, no visible folders, or fancy picture graphics, nothing but a dark screen with the white prompt waiting for you to type a command.

It was pretty boring to most people; however, it has evolved like the internet.

Most of its flavours today have a desktop version with an excellent screen, and pictures, as well visible folders.

In fact, Linux became one of the most stable operating systems even today.

Furthermore, as of today, every Android mobile device is based on Linux. Amazon Kindle devices are also based on Linux.

Now, if you ask me, I believe these devices are easy enough to handle and operate.

Well, Linux has changed, and the now, two decades later, everyone is aware of its existence, even if it's called something different. It has not stopped there of course. Cisco Systems are also based on Linux, and believe me; those are best, most stable, and fastest routers today exist.

Those babies are letting us use the internet with the speed that we have today.

However, if you might be more familiar with home based routers such as Linksys, ACER, LG, Motorola, Dell, Toshiba, Nokia, Sony, Sharp, Alcatel, Blackberry, Nexus, HTC, Google Glass, Archos, Samsung, Raspberry Pie, Pandora… well, they are all based on Linux operating system.

Besides cell phones and routers, there is other use of Android based systems too, such as in-car entertainment systems, Airplane entertainment systems, home theatre, gaming industry, digital security, and yet other devices, that are not even based on Earth.

Some other devices that are certainly notable, uses android based operating systems are used by NASA, such as Space Stations, Satellites, flight computes, even rockets.

Back to Bitcoin evolution, and think about where will it land in the next decade, or two?

I am not able to predict the future, however, if we might face a problem that required a solution, I am pretty sure that we will figure it out.

Bitcoin is getting it's fame already, it's only a matter of time to be world wide implemented, and to take over the world in some form.

# Chapter 14 – Purchase Power

I often being asked this question: Keizer! Who should I ask for permission if I want to make a payment with Bitcoin?

So my answer is always the same: You do not need to ask anyone for any permission. You can buy and sell Bitcoin anytime you want. Day or night, in any country, all you need is an internet connection. Bitcoin is not regulated by any nation, as this is not a government currency.

You can buy Bitcoin, and become your bank, even if you are under aged. It is not like walking into a bank, and you must be at least 18 years old if you want to open an account. You can also buy Bitcoin even if you don't have a job.

When you go to the bank and asking for a bank account, they will ask you to bring proof from your workplace, stating your position, as well how many hours you work and how much is your wages. It is ridiculous, and no wonder why over 2 billion people have no bank account. Anyways, there is no permission required from anyone.

**Getting paid in Bitcoin**

In case you are a freelancer, and selling your services online, you might choose to get paid in Bitcoin. Again you don't need to ask for permission from anyone, and you can get paid in Bitcoin.

**What can you buy?**

Well, you can buy anything on the dark web—of course—I do not recommend that, as you might come across criminals who would try to steal or hack into your bitcoin wallet.

Some cyber criminals would even try to blackmail you. However, if you do not provide your details, you should be just fine.

Realistically, more services are excepting Bitcoin, such as Hotels, Restaurants, Coffee shops, even some takeaway shops are now offering payment method using Bitcoin.

Large retail companies are also accepting Bitcoin, such as Shopify, TigerDirect, and much more. To see how broad range it can be, you have to look around where you live. The big cities have all sort of offerings, such as:

- Theatre
- Taxi Service
- Bicycle rent
- Private Jets
- Pubs


Also, you may consider other large companies that are now accepting Bitcoin, such as:

- Dell
- Microsoft
- Zynga
- Reddit
- Wordpress.com
- Subway
- Expedia.com
- Virgin Galactic
- OK Cupid
- Stream
- Alza
- Lionsgate Films
- Badoo… And much more

By using Gift cards, multiple applications also allow customers to purchase on websites, such as:

- Amazon.com
- Walmart
- Target
- Nike
- GAP
- BEBE
- Sears
- Papa Johns
- Best Buy
- iTunes
- eBay
- Starbucks
- Zappos
- CVS Pharmacy
- The HOME depot... And much more.

I wanted you to know that some of the largest companies already adapting to the idea of accepting Bitcoin. Moreover, to understand the range of goods and services that can be purchased, please see the list of categories that you may choose from:

- Airline
- Automotive
- Beauty
- Clothing

- Department Stores
- e-Commerce
- Electronics
- Gas
- Gifts and Toys
- Grocery
- Health
- Home and Garden
- Home improvement
- Hotel
- Jewelry
- Movies
- Pets
- Restaurants
- Shoes
- Sporting goods

As you see, the categories of shopping options are keep on growing, and if you are more interested in what stores you can pay using Bitcoin, you might check what you have nearby you, or what online platforms can deliver to your area.

# Chapter 15 – Get ready for the revolution

Why would I say that? – And what revolution? Well, you are reading this book; therefore you have an idea about Bitcoin; however, you can do a little test and see for yourself, where the rest of the world stands right now.

For example, I assume that you have some regular places that you often visit, even if not every day, but weekly or monthly. What you can do, is pick five or ten of them places where you typically purchase thinks like coffee, bread or milk, and when you just about to make your regular payment, ask the cashier if they accept Bitcoin.

I bet there are very few people out there, who would understand what you are on about. It is up to you, but you would be surprised how many people never even heard of Bitcoin or the word cryptocurrency.

I often ask shop owners if they accept Bitcoin, and see how they react, and I have to say most people are laughing, then say politely:, nooo, not yet.'' With a smiley face. I believe these are the individuals who have heard of Bitcoin previously; however, they have insufficient knowledge if there is any.

Most times people looking at me and ask me:,, What? What's that? NO!'' meaning they have no clue, and not even interested what that is. In fact, they might think that I am from those funny, candid camera tv shows, so probably they just want me to pay and get out.

I respect the Blockchain technology behind Bitcoin, however I don't mean to pressure people to get into Bitcoin, especially to invest all your money; however, it is time to get familiar with it. In my next book, I will explain how to buy safely, as well how much should you invest as a beginner, and how to keep your Bitcoin safe against hackers.

**Are we there yet?**

I will be sincere here, and I tell you right now, that we are not there yet. Unfortunately not yet, in fact, we are looking at ten to fifteen years, but it might be two more decades.

I only wish if that's a lie, but I have to be realistic and understand where do we stand right now. First of all, the technology that we have is great, and we already have it all that required implementing Bitcoin everywhere; however, changes take time. There is an enormous amount of work needs to be done to get those changes completed.

The reality is that Blockchain start-ups don't need to implement changes; however, there are so many existing companies that are just so large to make all required changes in all their systems, and to do so, would probably take five to ten years.

People are not educated enough, Blockchain developers need years of studying, and to get ready to work, everyday learning is essential for the rest of their career.

It is true that currently, we haven't got enough man power to carry out such changes.

So, basically there is not enough engineers yet, not enough software developers yet, those who could setup, as well maintain the system.

And again, the worse is that people want to become millionaires by investing in technologies, like Bitcoin or Blockchain, but when it comes to studying it for years, than only few people to participate.

## Marketplace

As I mentioned before, I guarantee you that those ten people you are going to ask about Bitcoin will have no clue, even they might have heard about it, they are probably don't have any, don't know how to buy or how to get paid in Bitcoin.

What it tells you is the market simply not ready for the change. Some places don't even accept credit cards yet.

Credit cards are not a new invention, yet many places still don't take it as a paying method. Some other places do accept it, but you have to spend at least five, some other places at least ten dollars, otherwise have to pay with cash.

## Nation states

If Bitcoin would be implemented everywhere in the world, there would be no more US dollar or Canadian Dollar, neither Japanese yen, therefore once all those would disappear, everyone would have to start thinking differently. Are we ready for that yet? I don't think so, but it's only my opinion.

It will take some time to adapt, but eventually will happen.

**Get involved**

Finally, I would like to close this chapter with good news; actually, it's excellent news! If you are getting familiar now with Bitcoin, there is no better time for sure! Faster you get involved and learn about the future of money, more comfortable your future will be. This is a fact.

Same thing, as everyone has to find out how to open a bank account at some point, in the future, everyone has to learn how to open a Bitcoin account, and to be honest, opening a Bitcoin account way easier than opening a bank account! Blockchain does not care if you are under age, neither your sex, or occupation.

Blockchain will not ask your wages, or for proof of address, it does not care your race, neither how you look like, in fact, it does not care if you are a human or not. Yes, that's right, in the future, machines can have a Bitcoin account on the Blockchain. Furthermore, they will be able to communicate with each other, and make transactions that you do today. For example, your fridge will know exactly when it will run out of Orange Juice, and it will be able to order it from Amazon, than an Amazon drone will be able to deliver your orange juice in no time! Once again, get involved in the future of money, talk to people who already involved, read books, and educate yourself. Understand how Blockchain works, and see how powerful it is, so that will give you more confidence for better understanding.

In case you are not interested in M2M aka machine to machine learning, that's fine, however, there are other reasons too, to get involved. If you look at Chapter 1 in this book, you might realize that on the 4th of August 2017, the value of Bitcoin was $2891,79; however just last night Bitcoin has reached the 4000 dollar mark, in fact it has hit all time highest mark of 4208.39 dollars.

These details are from coinmarketcap.com for your reference. Mainly in the last nine days, we have experienced around 40% growth. Meaning, if you have invested in Bitcoin, an amount pf 500 dollars two weeks ago, today that would worth around 700 dollars.

Now if you are a trader and you would invest $50K, within two weeks you could have made over $20K without even to be obligated to pay tax on it. There are millions, if not billions of people who don't make $20 even within a year, that's alone, should be just enough proof what a single cryptocurrency can do, more specifically Bitcoin. I am sure that you recognize the muscle of Bitcoin now, as well what path is going to, and indeed, hope that you have begun to like this young, yet super influential cryptocurrency!

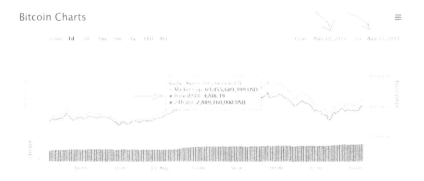

# Bonus Chapter – How to get paid in Bitcoin

Would you like to get paid in Cryptocurrency from your employer?

Never better to start than now! There is a company called Bitwage, that are now able to help you to get paid in Bitcoin, directly from the company that you are working for.

Bitwage is an international wage payment and staffing solution, that built on top of the Bitcoin Blockchain.

**Payments:**

Paying a domestic employee is straightforward. All you need is an account number, and you can start paying an employee. However, when start making international payments, that's when it becomes a little complicated. Unfortunately, many banks don't allow international payments to be made online. Instead, you have to confirm your payment over the phone, or worse in person.

When the workers are receiving the money they have worked for; they have to deal with unfavourable exchange rates, due to the money have been moved through international systems. As a result, we can see and experience, that freelancers are changing is higher, or even rejecting clients that are international, just because of these issues.

Imagine there are two different banks, that one resides in Bangladesh and the other in the United states. Each of them has their ledger system to keep their records up to date. However, there are some issues between the banks once they begin to interact. Because these banks have private ledgers, they have to work out between themself how they are going to send 1000 dollars from one to another. Back in the 1800-s, they had to put gold on the boot and ship it through the Atlantic Ocean; however, we are lot more efficient than that.

We have the internet, and many people believe that the current banking system is allowing easily to proceed with the international payment transfers; however, it is not true. Because both banks have a private ledger system, there is an issue of trust. To bridge these trust issues, the corresponding banking system was created.

Banks have build relationships with other banks they trust, by doing so, they have to develop corresponding accounts to each other. Because of different rules in different countries, they have built a chain of banks. So what is happening is that in Bangladesh a small bank first moves money to a big bank in Bangladesh, then to another large bank to the US. Then this large bank in the US will transfer to the small bank to it's original destination. However, if there is a currency conversion in the process, there will be even more intermediaries involved.

With each intermediary, there is a cost and delay, that is around 3-5 days on average, as well a cost on a currency exchange that is an average of 8%.

All these efficiencies have to do with the fact that there is a lack of trust in a private ledger system. However, with Bitcoin, thinks are indeed completely different.

From a high level, the Blockchain is a public ledger, that can maintain high level of privacy. So using this system, an e-mail can confirm the transaction, by having a link in the e-mail.

That link, is access to a highly secured public ledger, that undeniably proves that the money was sent from the small bank in Bangladesh to another small bank to the US.

Therefore, we no longer need any of those intermediaries, who approves the process, as well delaying the transactions and taking a fee.

As a result, international payments and Bitcoin are together, making payments across borders ever closer to the speed and ease to the domestic payment.

While this is happening, foreign remote workers or companies can maintain relationships a lot easier than before.

**Reputation of an employee or employer**

Employees know they must have an attractive profile, which provides good reputation to get quality work.

However, reputation can be easily manipulated by hacking profiles or faking CV-s. Using the Blockchain, you can achieve a reputation system that can not be faked.

As it turns out, how much you get paid, for how long and by who, are all objective statistics that are good for your reputation in an object of manner.

For example, if you are an HR recruiter, working for a company for year in full time, then another year for part time, that's value. Or if you are a network engineer who has been having a raise in every year in the past for years, that's another value.

All these information is valuable reputation. What turns out is that all these information is recorded on a Bitcoin transaction.

Therefore by receiving wages through a Bitcoin mechanism, you can have forever your reputation recorded on the Blockchain, in a decentralized way, from any client who pays you.

This is already happening, and remote workers are already receiving wages to obtain a Blockchain reputation. Unlike a subjective system that ends up producing lower quality workers, however they introduce themselves as high quality, using this payment status, it is indeed tough to manipulate the reputation mechanism.

This will help companies as well to trust in the reputation mechanism, and as a result, you will get a very efficient market between remote workers and enterprises. Bitcoin and Blockchain have been bridging these efficiencies and continued to do so.

In the future, companies will be made of decentralized work forces, which are glued together. Corporations and workers alike, will be able to find the perfect fit with one another, through objective payment reputation, with no differentiation between international and domestic payments.

Local workers will be empowered all over the world, while companies can leverage the competitive advantages of a diverse global workforce. Workers will be able to receive their wages faster and cheaper, while those payments can act as a reputation mechanism to help them find their next job.

If you like the idea of the future of payroll, either as an employee or as an employer, you might want to take a look at bitwages.com and find out more.

## Conclusion

Thank you for purchasing this book. I hope this title has provided some insights into what is really behind the curtains when it comes to the future of money.

I have tried to favour every reader by avoiding technical terms on how Bitcoin works that currently flooding the market worldwide.

However, as I mentioned few times, to fully understand how Bitcoin works, you may choose to read my books on Blockchain:

***Volume 1 – Blockchain – Beginners Guide***
***Volume 2 – Blockchain – Advanced Guide***

Volume 2 is very technical, however, tried my best to use everyday English, and making sure that everyone can understand each of the technologies and their importance, and how they are combined.

My upcoming book on Bitcoin will provide more details on how to invest safely, and what kind of wallets you required before purchasing any Cryptocurrency.

I will also provide guidance, on how you can become a miner by renting equipment, as well how you can start mining digital money using your laptop, or even your Android phone.

Lastly, if you enjoyed the book, please take some time to share your thoughts and post a review. It would be highly appreciated!

# BITCOIN
# Invest in Digital Gold

# Book 2

*by*
# Keizer Söze

Copyright
All rights reserved. No part of this book may be reproduced in any form or by any electronic, print or mechanical means, including information storage and retrieval systems, without permission in writing from the publisher.

Copyright © 2017 Keizer Söze

## Disclaimer

This Book is produced with the goal of providing information that is as accurate and reliable as possible. Regardless, purchasing this Book can be seen as consent to the fact that both the publisher and the author of this book are in no way experts on the topics discussed within and that any recommendations or suggestions that are made herein are for entertainment purposes only.

Professionals should be consulted as needed before undertaking any of the action endorsed herein.
Under no circumstances will any legal responsibility or blame be held against the publisher for any reparation, damages, or monetary loss due to the information herein, either directly or indirectly.

This declaration is deemed fair and valid by both the American Bar Association and the Committee of Publishers Association and is legally binding throughout the United States.

The information in the following pages is broadly considered to be a truthful and accurate account of facts and as such any inattention, use or misuse of the information in question by the reader will render any resulting actions solely under their purview. There are no scenarios in which the publisher or the original author of this work can be in any fashion deemed liable for any hardship or damages that may befall the reader or anyone else after undertaking information described herein.

Additionally, the information in the following pages is intended only for informational purposes and should thus be thought of as universal. As befitting its nature, it is presented without assurance regarding its prolonged validity or interim quality. Trademarks that are mentioned are done without written consent and can in no way be considered an endorsement from the trademark holder.

# Introduction

Congratulations on purchasing this book and thank you for doing so.

This book is an excellent beginner's guide to understanding what Bitcoin has to offer, and how it changes the future of doing business. The contents avoid technical details to provide a better understanding to those, who are new to this technology. There are certain terms that some technical background in Information Technology would help. However, it's not necessary. Everyday English has been used through this book to avoid confusion.

This book will take you by the hand, and show you step-by-step, how digital currency was born by analyzing historical data. For better understanding, first, this books is beginning with a brief introduction of why to consider Bitcoin at the first place, and why often called as Digital Gold. Furthermore, it takes a look at why is much likely, that the value of Bitcoin will hit 1 million dollars.

Next, we take a look at some examples and understand why Bitcoin is at its early days, and why it's not late to start investing. Following by wallet technology overview, explaining the key differences between hot wallets, cold wallets, and how to find out what suits you best. Next, I will recommend the best hot wallets, as well the best cold wallets that exist on the market, following by providing you all details on where and how to purchase them.

The book will carry on explaining Bitcoin ATMs, their purpose, how to locate them, as well how to use them for both: buying and selling Bitcoin anytime, securely, and offline.

Next, I will introduce the best online cryptocurrency trading platforms, and explain the key differences, as well their pros & cons. Next, will enclose how to recognize online scammers, and teach you how not to be fooled by online thieves. Finally, I will explain why it's profitable to trade with Bitcoin, and explain how to become a mixture of both investor and trader.

In the Bonus Chapter, I will provide some basics of Bitcoin mining, and its history, then finishing off on how you can mine Bitcoin using your old laptop, or even your Android phone with a free online software.

There are plenty of books on this subject in the market, thanks again for choosing this one! Every effort was made to ensure the book is riddled with as much useful information as possible. Please enjoy!

## Chapter 1 - Why consider Bitcoin

For the first time in human existence, we have a secondary option to a none governmental controlled currency. This is awesome! This is when you and I, in fact, anyone can participate. This, itself is why the most important, however, let me expand on it with further detail.

There are many cryptocurrencies out there; however Bitcoin was the first amongst them. Cryptocurrency works, and it's simply put, provides additional options. There are multiple opportunities time to time that could potentially benefit you and your family. However Bitcoin is one of the best options, and I will tell you exactly why. If you look at the current banking system and understand how they manipulate the inflation rates and lending out money that hard working people saved over the years, you know that is absolutely ridiculous.

For example, I am saving 1000 dollars over the years, that I will give to the bank, so they can give me 1% interests after a year. Of course not allowed to touch it. However, if I want to use it, not only all my interests would be lost, but I would be even charged for it.

So, then I choose to keep that money in the bank, then you would go to the bank and borrow that money, however, once you would do that, your interest rate to pay it back would be 20%, or even more.

Now, this is over the limit. Don't get me wrong, and I do understand banks have bills to pay too, as well hard working employees, security, computers, antiviruses, firewalls, and all; however, the option remains.

In 2008 when there was a largest financial breakdown in history, banks have decided to provide money, as a bailout for insurance companies, car companies. They have began printing paper money like crazy, causing devaluation of all those existing hard working earned money. Therefore it has devalued the economy too, and increasingly created inflation. Once the interest rate is getting higher, you have to think about surviving.

In case you have lot's of money in the bank, and tomorrow we are going to hit the next financial breakdown, the banks can decide to not giving you any money. In fact, they can call in bankruptcy, meaning, you will never see your money again.

It has happened over the last 50 years in multiple countries, and people, as well small businesses have become poor, or worse, homeless. Financial crisis will happen again, and it will happen more often than before. Firstly, because the last one back in 2008 has not dealt with, all have happened is a workaround, but no real fix.

Secondly, the world debt has only grown since, and soon will be time for demanding those debts. The problem will be the same as last time or the time before.

The debt will not be paid by the banks, who are responsible for creating the debt in the first place, but the bank will demand from those who have mortgages or other interests to pay back to the banks.

To think otherwise would be certainly unwise. The money that banks have, well is a bit of funny money; indeed it's kind a believable money.

Those times long gone when gold or silver was backing money up, those were real money, however, nowadays it's only numbers, or digits, that the banks told us that is worth something. Meaning worth nothing.

What Bitcoin provides us is a revolutionary currency that not controlled by the government, neither any bank or the Federal reserve system, therefore providing us an additional option to FIAT money. You can buy Bitcoin using multiple different mobile phone based application for free.

You can also use the desktop version of applications, or even individual exchanges, or even from local Bitcoin meet ups. Additionally, nearly any currency is exchangeable to Bitcoin. Therefore there is no excuse.

Another thing is that Bitcoin is not just some sort Day trading option, something that you can invest only. However, you can also earn Bitcoin. I have explained in my previous book:

*Bitcoin Blueprint - Volume 1*

You can receive Bitcoin, using the company called Bitwage, while it doesn't matter what company you are working for. Besides to Bitwage, there are other freelancing companies too, where you can provide your services in exchange for Bitcoin.

Some of those online freelancing platforms are Upwork, FIVERR; however there are much more out there, and with a simple google search you find much more. So once you have Bitcoin, you can make a transaction to any destination in the world.

Imagine that you want to send only 100 dollars to China, and you have to deal with high bank changes, as well currency exchange charges, not to mention the minimum five days delays.

Of course you might use PayPal or other trusted third party services; however the transfer delay could take even longer, and the fees are also will become higher, which are at least 5%.

In the other hand, transferring Bitcoin, the transfer charges are on average of 0.5% to 1%, independent of how much value of Bitcoin you are about to transfer, additionally the transfer will be immediate, and in a maximum of 10 minutes, it will be validated on the blockchain forever. As you can see, when using Bitcoin or any other cryptocurrency,

## YOU ARE THE BANK.

You have to understand that are no other intermediaries required for you to make a transfer to anywhere in the world.

The only thing is needed really is to have a Bitcoin account for the person that you are to transfer to. Bitcoin account can be downloaded to a cell phone in less than 2 minutes. There is a misconception where people believe that once you lose your phone, your Bitcoin will be lost too. This is not true.

You can back up your phone, and even you would lose your phone or break it, using another device you can log in to your Bitcoin account anytime, from anywhere in the world and access your account.

As soon as you recall your account on the new device, you will be able to make additional transfers or exchanges to different Fiat currencies, or even cryptocurrencies, all you need is internet access.

As I mentioned before, finally we have access to purchase Bitcoin, or any other cryptocurrency using our phone, and this is from our own will, instead of asking for appointments from the bank to open bank accounts, and all those annoying procedures.

Additionally, it has been observed that each time when a nation's fiat currency devalues, Bitcoin price increases.

This is easily explained by those who affected, learned that hard working earned money can be lost because of government manipulations, and it's time to put savings into Bitcoin.

The worse of all is that more than 2 billion people around the world have no bank account, and those people might be not allowed to have an account due to the banks; however, some others may just don't want to participate as a bank account holder.

We all can have a Bitcoin account, and all we need is a phone, or even if we can not afford a phone, all we need is internet access when we wish to make a transfer.

This is opening another new market, as those who were afraid of local government or local mafia, they can simply have all their money in Bitcoin, registered on the blockchain.

Other noted accomplishments, such as charities that are also corrupted, now can freely receive the exact amount of money that you or I would intend them to receive.

It is very similar to crowdfunding, and the concepts are simple. If I want to give 100 dollars to a charity, how come they are only receiving 30% of the sums of money?

Well, the rest of the money goes to the currency exchanges, banks fees, marketing specialists, and whatever people who are part of the setup in the first place.

Instead, I can transfer the exact amount of money as person to person, or person to business manner, and I will know exactly how much they will receive as the transaction will be recorded on the blockchain.

Furthermore, speed of implementation. Imagine that someone requires 1000 dollars as soon as possible, and you choose a bank for transferring.

Imagine that the money only will reach the destination, and become withdrawable in the next five days.

That's crazy slow, especially in 2017, is just utterly ridiculous!

Imagine that five days later the money would be too late because someone would require something urgently, the good will wouldn't be enough if you choose the bank.

Then again, in the other hand using Bitcoin, the transfer is immediate and gets validated within a maximum of ten minutes from the time of transfer.

Not enough reason to consider Bitcoin? Then let's dive in!

## Chapter 2 – Peer-to-Peer Economy

Yes, you have heard of Bitcoin, as well Ethereum, Monero, Litecoin, and they are indeed are cryptocurrencies. What you have to understand, is they are all functioning on the blockchain. Without the blockchain platform, they would not exist.

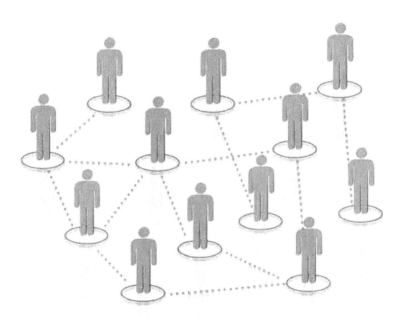

Therefore, firstly you have to understand what Satoshi Nakamoto has done with Bitcoin. He was able to solve the problem of manipulating code. Before his invention, nothing was stopping you to copying the same digits and sending to multiple destinations. Having created the blockchain, Satoshi was able to solve the problem. This, has now changed the world of posting digital code, so let's look at an example.

If I were to send you 100 dollars worth of Bitcoin from my account, I would have 100 dollars less of my code. Then you would send that code to your friend, you wouldn't have it anymore, as your friend would have it, and so on...

However before the blockchain, it was impossible, in fact, people have been working on this since the late 70's, yet no one was able to solve this issue. Against the centralized banking system, Bitcoin transactions are now empowering the end-users by becoming the bank itself.

So basically there is no inflation that the banks or governments can manipulate, neither controlled currency of what I can or I can not use, and most certainly no interests rates. With Bitcoin, there is no quantitative easing, simply because it has a pre set algorithm that set's it to 21 million Bitcoins.

There will be only 21 million Bitcoin mined ever. Therefore you can't print more Bitcoin, neither can control the inflation, nor the interests rate. Bitcoin is a real decentralized peer-to-peer system that can store v

alue. Those, who unbanked, now may become their bank, and more people using Bitcoin all over the world, and more network effect will be; therefore those people can create an even better peer-to-peer economy.

**Freedom of choice**

What freedom of choice does, is that you can send Bitcoin all over the world by using your account, instead of your name. Even though your address will be registered on the blockchain once the transaction is validated, it's done, however, your name will not be there.

When you think about it, anonymously sending money from anywhere to anywhere, to anyone, at any time, it is power. The reality is the majority of the world is living in a particular government society, and they do not have the same freedom as others.

Therefore it is complicated for them to speak up, and this is why you can see a huge rise of Bitcoin in countries such as India, where there is a currency war right now. As you see, some of those people can not speak up, they are fed up with the government, as well their money track down system, and now they have a choice to opt out into a Cryptocurrency such as Bitcoin.

Opt out to currency, that is own by a peer-to-peer network, where they can take control of their hard worked earned money.

Again, what Bitcoin-Blockchain system does, is designed in a certain way, where it doesn't matter where you live, or what race you have, even what is your religion or political view, simply because anyone can join the global peer-to-peer system.

Bitcoin creates a digital identity, instead of providing your name, in a decentralized manner.

# Chapter 3 - Is Bitcoin dead?

Short answer: yes. Bitcoin is dead, in fact, Bitcoin is so dead that poor Bitcoin has died 151 times already! ...yet it still exists!

If you are thinking how is that even possible, then let me elaborate on this topic with further details.

Most times when Bitcoin makes it to the news, often have some negativity. Bitcoin is most famous of being a criminal currency and thinks like that. Well, Bitcoin can be transferred all over the world with anonymity; therefore criminals love to use it.

Should we burn it to the ground then? We should right?

The reality is that criminals use cars too. Should we burn the cars as well?

Oh no, we can't, because someone would lose out on profit. How about cell phones? What I understand, is that criminals use cell phones too! We must stop producing mobile phones too. But you know what? I heard that criminals using the internet, and Facebook too! So we have no choice, but to kill the internet and facebook and twitter, (especially twitter). Common now!

Or, is it just about the currency? Then again, criminals do use hard cash, as well have bank accounts and making transfers too, some of them also paying tax, and bills, but anyway... Like everything else in the world, can be used for good as well for bad intensions.

Computers can be utilized for studying as well hacking. Police officers have guns, and criminals have guns too. Michelin chefs have very sharp knives, and some underground thugs have even sharper knives too.

We can't just stop producing everything in the world! Wake up! Not you, but those who apparently believe that Bitcoin is only for criminal use. This is ridiculous, however many people still think that Bitcoin is a money of the criminals like rapists, drug dealers, arms dealers, terrorists and all the evil that exists.

I am not saying they don't use it, yes they certainly do, but Bitcoin can change the life of billions, and that will be appreciated and should be focused on. Criminals will be criminals anyway, but please stop spreading nonsense about something that is a truly revolutionary technology. Unfortunately, Bitcoin has been introduced with a bad reputation, and this has triggered to create even more fake news over the years.

Another one is when Bitcoin hits a milestone in value. Bitcoin was dead when it has reached $100, then died again when it hit $200, then again when it hit $1000, and so on and so forth... However the value of one Bitcoin today is $4137.33. Today is 18th of August 2017.

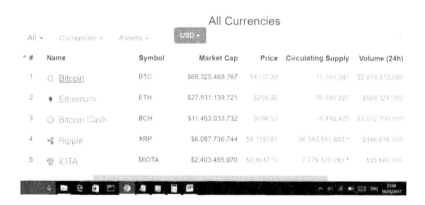

There is a website that is specializing in; Bitcoin is Dead."
This site is a collection of all those news where Bitcoin was in the news as a dead currency for

all sort of reasons. If you can click on the following link, it will take you right there:

https://99bitcoins.com/bitcoinobituaries/

First time when Bitcoin was dead, it was in December 2010. Since that painful event, Bitcoin has died another 150 times, yet still, exist. So what we can learn from it, is more and more Bitcoin will die, more it will hit the news. Therefore more people will hear those fake news when Bitcoin has died.

When you think about it, imagine that you heard in 2010 that Bitcoin is dead, then you read another article in 2011 where Bitcoin is dead again, but then in 2012 you will hear it in the telly that Bitcoin is dead, in fact, Bitcoin has died in 2013, 17 times. So, obviously something isn't right. Anyways, I wanted to share with you this link and another fake news: Bitcoin is dead! This time dead really, because in 2017 until middle of August it has died 32 times already.

Ok, maybe you not going to believe me, so I will tell you the truth instead: Bitcoin will carry on end up dead again and again in the news.
However,

**...Bitcoin is here to stay!**

## Chapter 4 - Will Bitcoin hit $1million?

Now, this is a more interesting question, and again the short answer is: yes. Yes, Bitcoin will hit the value of 1 million dollars for sure; however, the real question should sound like this: When will Bitcoin hit a $1million mark?

As you see the real question is the date, instead of: does Bitcoin has a capability or not.

There is a huge industry around Bitcoin, and people already made a fortune of investing into it.

Due to Bitcoin's value increasing over the years from cents to dollars, then from hundreds of dollars to thousand dollars, and only this year from one thousand to two thousand, then all the way up to 4500 dollars, it is possible to see Bitcoin reach 1 Million dollars.

There are analysis, where predictions of Bitcoin can capture between one and ten percent of the global forex market. That implies that the price of Bitcoin could rise between 100,000 to a million dollars.

Most of the people who on the sidelines and not buying Bitcoin today will start to buy Bitcoin when it reaches 5000 dollars. The greater percentage of the people will invest into Bitcoin when it starts to get over ten thousand.

Some predictions already stated that Bitcoin value would reach 10000 dollars by 2010.

There will be ups and down like any other technology, however as I mentioned earlier, Bitcoin is here to stay, and some other notable people like the CEO of Zappos has been predicted that Bitcoin can reach the value of between 500,000 to a million dollars by the year of 2025.

Most people only see the charts, and thinking how high it can go, however, if you begin to count backward instead of looking at the charts, you may look at it in another way, which is this:

This is a transactional currency, and It is a store of value. A such, like it, is a product and a service competing on a very substantial market for storing value and transactional currency.

Therefore, you should look at the size of market of that, as well how big the market share, then you can start calculating what Bitcoin can realistically take in a realistic timeframe.

Once you ask that question, then you will come up with the market capitalization of Bitcoin total.

Next, you divide that, by the number of Bitcoin in circulation by the estimated time, then you can come up with some between two million to five million dollars per Bitcoin.

My suggestion is this: if you buy at least one Bitcoin as soon as possible, that could be just enough for you to become a millionaire in the future.

Bitcoin is a currency of the future, no doubt of that, however you have to look at it on a bigger scale and understand that Bitcoin itself would be worthless, however, because it's scaling on the blockchain technology, it will become not only a currency, but a payment system.

This is because of the blockchain encryption, that has a greater ability to bring more of the world of the population out of poverty than anything else we have ever experienced.

It is truly one of the most important inventions of the history of human kind, and certainly most important invention since the internet.

This is going to improve the lives of every single person on the planet.

If Bitcoin will become a real money and excepted worldwide in the next ten years, it can become easily worth of 1 million dollars.

If you look at the value of US dollars in circulation, that might be between 5-15 trillion dollars, then divide it by the total amount of Bitcoin in circulation that is currently 16 million.

Next, if you use even the higher value of dollars calculated, that is 15 trillion, next let's calculate how many Bitcoin will be in circulation within ten years, and that is around 18 million, all is needed is to divide that figure, and it comes out to 833,333 dollars.

Well not exactly 1 million dollars, but it is a very close digit. Of course, Bitcoin might become only mainstream within 20 years.

However, due to our current technology and having the internet, we have learned, that this process can become quicker.

In fact, there are researches shows, to accomplish something that is related to technology as well global payment systems using the internet, we can speed up the process in average of 3 times faster.

If that figure is accurate, that could also mean that Bitcoin can be accepted worldwide within seven years.

If you would calculate the seven years, instead of the ten, using the same formula as I have just exhibited earlier, it could mean that Bitcoin can even expand beyond 1 million per coin.

## Chapter 5 – Are you late for Bitcoin?

Should you invest in Bitcoin?

First of all, if you are asking yourself a question, it might be a wrong question to begin with. The question you should ask is:, when will you invest in Bitcoin" However let elaborate on this topic.

The time will come where financial systems will collapse, or people will lose faith in their government, or worse, when the inflation makes you homeless.

Those times will be when Bitcoin should be looked at, as a vehicle to invest in it. However, you may choose to see Bitcoin not as an investment, but a new opportunity. If you start to think differently, it might become more apparent to you, that is not just can become the best investment you would ever do, but merely a new opportunity to try out something that you have never experienced before.

You might aim to become a Bitcoin millionaire; however, you don't have to, at least not at first. Instead understand first what Bitcoin can provide to you, as well to everyone on this planet. If you look at those who already become a Bitcoin millionaire, you might realize that there is a pattern between them.

That model comes down to faith in Bitcoin, as well a full awareness of what exactly Bitcoin is. So first of all, you might be overwhelmed like one of my friend Baggio, who want to invest 1000 Euro, but he hasn't even got a wallet yet, in fact, he has no clue what the wallet is in the first place.

There are people, who have saved of 10-20K or even more and bought Bitcoin with it, then some of them find themselves unable to access their wallet, hence all money lost in the cloud.

Instead, buy today for 50 dollars, but at first definitely no more than 100 dollars worth of Bitcoin, and start to understand how it works.

Make a payment first, you might buy something on Fiverr, or Upwork, or even purchase something on Overstock.com or some of the Shopify stores, either way, my point is, to make a payment with Bitcoin.

Then you might try out some of the exchanges and see what best works out for you, and what platform you prefer. My point here is this: do not think that you become rich overnight, or think of it as a fast instant cash, and it is because Bitcoin doesn't work like that.

As I mentioned a few times already, Bitcoin is here to stay. Therefore you have to look at it in the long term, as it will influence our lives in the future.

Again, the point is not how much it will worth and when, instead, take this opportunity, by start putting your toe in the water, and begin to understand the system that we are entering. It will transform everything that is currently part of our life, therefore it is a good idea to have some basic knowledge, as a minimum.

Society is changing rapidly, especially with the technology that we are using today, and if you want to be ahead of the game, and want to understand where the human society is heading, such as automation or machine learning using Blockchain, it is certainly mind-blowing what the future will bring us. A future that is already in the testing environment, and the stable root, Blockchain already in place. Those that are not interested right now, it's ok, however those people will not read this book anyways.

Those that are not interested today, one day, that will come soon by the way, will realize that it might have been better to get involved a bit earlier. However, if you have no clue what Bitcoin is or where it's heading, and this is your first book on Bitcoin, and you already made it to this chapter, then you already involved!

When it comes to a second option of transferring money on a medium, most people think and might choose another bank, Paypal or Payoneer, but you have another option. Or at last you are aware of it, and that itself can help out many people.

Bitcoin has been helping out millions of people already in the last few years, and it's only warming up. Of course, at the same time, getting stronger in value, as well in good reputation, and getting more accepted in many other countries too.

# Chapter 6 – 11 Reasons to invest in Bitcoin

I have explained many reasons why you should invest in Bitcoin; however this time I will try to summarize them for you to have a better reference guide when it comes to this question:

**What are the possible reasons that you should invest in Bitcoin?
Here it comes, so get ready!**

1. Solves multiple issues: You don't need a bank, anyone can have a wallet, and it's free. You can send money to any person in the world at any time, with a little fee.

2. People in under developed countries, can have access to Bitcoin, and all it's required is Internet access

3. No central governance, therefore no one can control your money.

4. Optional: Nobody forces you to have Bitcoin. All through most countries, you must have your home currency when you go shopping; however, you can have Bitcoin too, in any country that you live in.

5. No single point of failure due to its peer-to-peer network

6. No Inflation, basically there is no continuous injection of Bitcoin into the marketplace, due to its fixed supply to 21 million.

7. Growing demand: while there is a rise in demand, the value per Bitcoin will continuously increase

8. Legalization of Bitcoin: there are more and more countries where Bitcoin already legalized such as US, UK, Australia, New Zealand, Netherland, Ireland, France, Belgium, Spain, Portugal, Malta, Greece, Bulgaria, Italy, Sweden, Russia, Norway, Iceland, Finland, Lithuania, Denmark, Estonia, Slovenia, Slovakia, Romania, Poland, Germany, Croatia, Czech Republic, Vietnam, Singapur, Thailand, Malaysia, Indonesia, Philippines, Taiwan, South Korea, Japan, Hong Kong, China, Pakistan, India, Lebanon, Jordan, Israel, Cyprus, Columbia, Chile, Brazil, Argentina, Nicaragua, Canada, Zimbabwe, South Africa, Nigeria.

**NOTE: This data is according to the date of August, 2017.**

9. Decentralized ledger system: Bitcoin is running on the technology called Blockchain,

that has become very famous in recent years for its security; therefore the system is fully trusted. Bitcoin was the first application of the blockchain technology.

10. Respected by influenced people: Bill Gates has stated that Bitcoin is better than the single currency, however amongst others, such as Richard Branson, or Warren Buffett, John McAfee, and much more. In fact, John believes so much in Bitcoin that he has predicted that by 2020 a value of Bitcoin will reach 500,000 dollars. Besides all the famous names, there are prominent companies as well who accept Bitcoin as a payment method on their platforms such as Microsoft, Expedia, Dell, Subway, Reddit, Steam, Alza, Virgin Galactic, and much more.

11. Early adopters. There is a graph called S curve that defines the adoption of technology by humans. This chart exhibits the types of people who and when they get involved in a particular technology in a percentage.

S curve has been studied before on an adoption of multiple technologies over the years, for such as Telephone, Electricity, Auto, Radio, Refrigerator, Stove, Washing machine, Dishwasher, Microwave, Colour TV, Air Conditioner, Computer, and Cell Phone

The S curve tells us, that in any community, there are five different types of people or adopter groups. These groups can be categorized by the time it takes them to adapt to a given innovation or new behavior. To identify what groups already approved the change and which groups have adopted to the innovation in different points in time. I describe them in order, so please find the adopter groups as below:

- Innovators > 2.5%
- Early Adopters > 13.5%
- Early majority > 34%
- Late majority > 34%
- Laggards > 16%

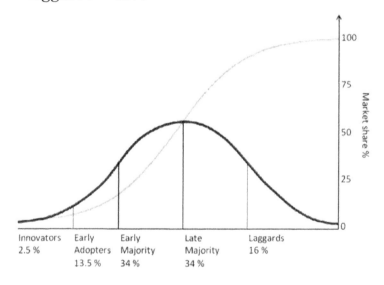

Now that's being said, Bitcoin exists since 8.5 years, and all 21million Bitcoin will be thoroughly mined by 2140.

Meaning the actual mining process will take around 131 years to fully complete, calculated since 2009.

That means, we are probably not Innovators anymore, however as of 2017, eight years on the run, and we have another 123 years left until all Bitcoin will be in circulation, I believe that we are in the age of Early Adopters.

However, if we only start to adopt Bitcoin in two decades later, we would be already in the Early Majority group.

Of course this is only my personal opinion; however, I also believe that those who will be born after 2070 or at least start to learn the economy at that time, will be already in the Late Majority group.

Even if they are eager to learn the technology fast, by then it will be the must, similarly to nowadays, where we all should know how to turn on a laptop.

If you can look at the consumption that spread lots faster today, still will take some time for people to adopt Bitcoin. For example, if you look at the adoption of the cell phones, it has reached 50% in 10 years, same with the internet.

However, the easiest way is to find out where we stand today, is by asking ten random people every day if they have Bitcoin or at least if they are aware of what that is.

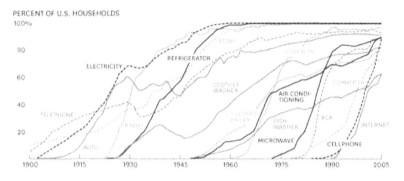

## Chapter 7 – Potential risk of Bitcoin

Let me begin by clearing one thing. Safest and easiest way to buy Bitcoin, is not always come together. Buying Bitcoin is easy, but you always should know exactly what the risks are.

Like I mentioned before, one of my best friends wanted to start investing to Bitcoin with 1000 Euros, so I asked him:

Where do you going to keep it? He replied: Can you buy it for me? – and look after for me, please.

Well, this is not the way to go about investing your money, let me tell you right now. So, because you heard that Bitcoin is the future, and it will make a millionaire and all that, which is true by the way, does not mean that you should jump into any buy without knowing the potential risks. Let me elaborate on this.

If you have a bank card and you forget the PIN code, you can call your bank, and they will help you, or at least they should, however when it comes to Bitcoin, there is no support line to call.

Or, for example, if you choose to type your bitcoin wallet address, and you mistype it when you purchase, and the Bitcoin will get transferred to the wrong address, that's another bad news for you.

You can not call Mr. Bitcoin at Bitcoin bank and ask for a refund because you made a mistake. The reason you can not call Mr Bitcoin is this:

There is no Mr. Bitcoin, neither Bitcoin Bank. It's a decentralized digital currency, where once transactions get validated, end of story.

If you are not scared yet, then let me tell you what happens when your wallet gets hacked.

I have a friend, or I should say an associate, who has ignored to buy a cold wallet and kept his Bitcoin on his mobile application, and to be precise, he had three Bitcoin a few months back when the value of Bitcoin was between 2500 – 2700 dollars.

Yes, he got hacked, without even realizing it, and the hackers ( or just one hacker) have taken all his Bitcoin, all of it is gone. So he asked me what to do.

Now let's be realistic here. You can't call the police, because they have no clue how to handle situations like that, and you can't call the Bitcoin bank either because there isn't any.

If you get hacked, and your Bitcoin disappears, you have to get on with it, and understand that it's is your fault, not the hackers.

Ok, that's not true, as it's the hackers fault too, but let me ask you one thing:

Do you leave your front door open? Or do you leave your car doors open? Maybe these examples are not close enough, but then let me ask you this:

Do you leave all your cash, dollars, pounds, euros whatever you have out to the public where anyone can grab them?

It is clearly an invitation for thieves to act upon right? For your sake, I hope you agree that your value should be secured, instead of open for anyone to grab it.

Well, if you didn't know about hackers, it's still no excuse, and still, no one will be able to help you, however, now that you know the potential risks, I am sure you agree that's better to learn how to secure your Digital currency.

In fact, step one of investing in anything, is to understand what are the potential risks.

There is no danger free business, and when it comes to Bitcoin, it's the same story.

By the way, if Bitcoin would be worthless, hackers wouldn't care about it whatsoever. However,

## Hackers LOVE Bitcoin!

**Hackers VS Bitcoin**

Hacking used to be fun, and playground amongst techies, however as the information age has changed, hackers have grown up too.

It might have been fun to break into a Company's system and leave a trademark, so hackers can become famous online, using their aliases; however, the game has changed.

Hackers have gained superpower over the recent decade by able to access any digital information.

Not only individuals, but hacker groups have formed, and large Cybercriminals were born.

Taking it further hackers realized that stealing passports, ID-s, bank accounts, you name it, then sell them on the dark web, is not only very risky, long process, but not very profitable either.

With the Bitcoin value in the continuous increase, there is a significant demand amongst people with no technical background, hackers have learned that, there is a new market for them: Stealing Bitcoin.

You see, they don't have to take the risk of selling passports to those who might be the FBI; instead, they can steal Bitcoin and other cryptocurrencies.

Then they can transfer funds anonymously, using TOR, or even exchange them to Fiat currencies offline, using a Bitcoin ATM-s.

This is not a joke, and believe me, once your Bitcoin account get hacked, and your funds are transferred, it's game over.

There is only one way to be secured against hackers, and that is to have a cold wallet, aka hardware wallet.

I will explain in more great detail on wallet technology, as well I will recommend the best wallets in some later chapters.

## Chapter 8 – LocalBitcoins

Before you run away and don't want to hear about Bitcoin anymore, let me tell you good news. You don't have to be a Chief Technology Officer to understand how to keep your Bitcoin safe.

In fact, you don't have to know about Elliptic Curve Cryptography or Discrete Logarithm Problem, or anything on how Bitcoin actually works, but you have to understand how to secure your Cryptocurrency.

Step two of investing should be to learn how to avoid potential risks by securing your investment. This brings me to another topic, and that is this:

### Be aware of scammers!

There are so many ways to purchase Bitcoin, and some people come across a technique of using local Bitcoins.

You can reach the website by using the link:

### https://localbitcoins.com/

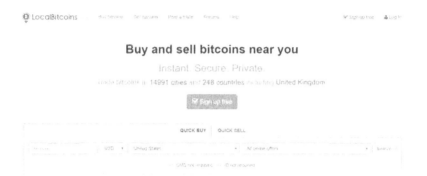

This is an attractive market where people trade cash for Bitcoin, or Bitcoin for cash; however, there are also scammers here.

Similarly, when you would buy a second hand item for example from e-Bay, you should be aware that the item might have been stolen. So, you would have to see if the seller has good feedback, ratings and so on. You have to understand that you are dealing with strangers.

So for example, if you see someone selling first time with no rating, your chances of being scammed are 50/50. What you have to understand is that if something is too good to be true, then probably is a scam.

If you see a new user on Local Bitcoins who sell at a lower rate than average people, if I were you, I wouldn't risk it, simple as that.

However, you can approach those users who are members of the site since at least 3-5 years.

That's when I would feel safe to buy, however, if someone you trust, recommends someone, that's also might be a right choice.

Those who have hundreds of reviews, you can read some of the recent ones, and decide it for yourself by understanding how trustworthy they are.

Some of these people might live locally to you, and you can meet them in person, then exchange real cash for Bitcoin, or vice versa.

Local Bitcoins is a good place to start buying Bitcoin for beginners due to its easy understanding. You can sign up for free, and the interface also has a forum, where people post questions and answers.

This is an excellent way to gain additional knowledge of the current market or anything that you have in mind.

Naturally, you can post a question, or you might go through existing threads and learn how people are talking to each other.

It's also good, as you might find someone who not only locally close to you, but you might know the person, as it has happened to another friend of mine.

Some people also post on the forum in different languages than English only, as well people share their experience with scammers as well.

However before you would buy Bitcoin from anyone, you must have a wallet.

## Chapter 9 – Hot wallets

My recommendation is to read this book and learn more before you buy any Bitcoin. However, It's time to understand the wallet technology. First, I would like to point out that there is a misconception, or I should say misunderstanding what the crypto wallet is.

Many people believe when purchasing Bitcoin, it will be stored on the wallet. This is not true. Bitcoin that you will purchase will be recorded on the blockchain, in fact, all Bitcoin that ever was mined, and will be mined, are always going to stay on the Blockchain.

Bitcoin has no physical presence, and when you are transferring Bitcoin, you don't move the Digital coins anywhere; instead, you assign a new wallet address to the amount of Bitcoin you buy or sell. Wallets are a software technology, which stores your private and public keys, and interacts with the Blockchain to allow you to access your Bitcoin account.

Once you have access to your account, you can send, receive, or directly monitor your balance. There are many different kinds of wallets, and some portfolios support different currencies, however, some only supports one particular crypto currency. For example, imagine that you have an HSBC bank card that you can access your Euros; however, you also have another card with Bank of America to obtain your dollars. When it comes to a crypto currency wallet, some wallets can support multiple currencies simultaneously.

**Online wallet**

You might choose to have a wallet from Blockchain.info that allows you to have access to your Bitcoin as well Ether. By the way I have a Blockchain wallet, in fact, that was my first wallet, and I still have it, even it's limited for currencies, I still use it, and recommend to anyone to start with it.

You can have a blockchain wallet for free on your desktop as well on your smart phone by reaching the website called: blockchain.info

Once you enter the site, you can click on the menu: wallet, then click on: Get started now. That will take you to the registration page, where you can sign up in no time, and buy Bitcoin very quickly.

You can link your current bank account and start to purchase your first Bitcoin. This is how I bought my first Bitcoin, and I would highly recommend it for beginners.

It's effortless to use it, and start using your first Bitcoin wallet. This wallet does not support all the currencies that you might want to invest in the future. However right now the focus is on purchasing Bitcoin, or even before that, to have your first crypto currency wallet that supports Bitcoin.

If you start with a Blockchain wallet, which I highly recommend, do not go crazy and invest all your money. Instead, have a feel for the technology, and start with 100 dollars. Some people recommend no more than 500 dollars as a first investment, however, if you are a complete beginner, you can do even just 50 dollars worth for the first time. When I have started, I went in with 100 dollars. When you link your bank account, the options to purchase Bitcoin for 50 dollars, the fees seemed to be higher than buying for 100 dollars worth, so I have chosen to have 100 dollars as my first investment.

As I mentioned before, easy does not come with safety, and it is also true with a Blockchain wallet. For instance, if Blockchain.info gets hacked, the hackers can take control all wallets that reside on their platform.

It had happened before with other trading companies, which resulted in administration; therefore you have to understand, that any online platform exists today is never 100% secured.

Also, once you have a wallet installed on your mobile phone or desktop computer, those devices also can be hacked.

I am not saying that once you buy Bitcoin, you will be hacked in the same day, but there is a potential risk, and you have to live with that.

Of course, if you back up your wallet, you have a better chance of being able to retrieve your funds, in case something bad happens.

**Desktop wallets**

Desktop wallets are reasonably secured; however, they can only be used on your desktop computer.

Once you have installed it on a particular computer or laptop, that will be your only device where you can access your Bitcoin or other cryptocurrencies.

As I mentioned before, it is not based in the cloud, and if you regularly turn off your pc, and having an antivirus, you should be just fine.

Of course, there are many ways for hackers to get hold of your desktop wallet using various phishing attacks, or viruses from drive by download sites or torrent websites.

Once the hackers would access your private keys, they would be able to access your digital currencies; therefore, it is not recommended for long term use, or even for short term for large sums of coins.

Again, desktop wallets are easy to be downloaded, as well to be used, however still not fully secured.

## Mobile wallets

There are many different mobile wallets, and Blockchain.info has one as well. However, Blockchain.info provides both, online and desktop wallets.

On the other hand, there are some mobile wallets, specifically for cell phones. Having a mobile wallet is essential for making payments anywhere you go, as your cell phone probably will be with you most of the times.

Mobile wallets can provide good security too; however, you must back up your mobile wallet too, same as your desktop wallet.

For example, if you would lose your phone or break it, buying another phone and having backup phrases for your mobile wallet, you can simply back up your new cell phone, like nothing happened.

Again, if you don't back up your mobile wallet, you can lose access to all your Bitcoin forever.

# Chapter 10 – Cold wallets

Cold wallets aka Hardware wallets, those two descriptions are often used both, and really means the same thing. One of these devices, is what you must have if you planning on purchasing large sums of Bitcoin.

These are the best and safest wallets that you can have. Hackers can't do much by attacking it from online, as the cold wallets are keeping the private keys on the hardware, away from the internet.

Some of these wallets, are often look like a USB Stick, and unfortunately, people believe that having a USB stick is the same as having a cold or hardware wallet. This is not true.

USB stick cannot be backed up; neither understands the current market. Of course, hardware wallets do look like USB devices. Hardware wallets also have to be connected to the internet for sending or receiving funds. However, this would require a special USB type of cable.

Any cryptocurrency that you are afraid to keep it on a mobile or online wallet, you should keep them on a cold wallet.

There is no way to someone hack your cold wallet, unless they have physical access to it. However, still they must break your pin code, then must have access to your secret keys that you write down on a piece of paper when you use the device first time.

The only downside, in my opinion, is that if you want to sell some of your Bitcoin quickly, you can not. I mean if you have your cold wallet with you, and there is a desktop pc around, then it can be done in less than 5 minutes.

However, when you have a mobile wallet, transaction can take like 10-15 seconds. Traders always have lot's of Bitcoin on their mobile wallets too in case it's time to sell or buy.

I call myself an investor, when it comes to Bitcoin or any other cryptocurrency. I do believe that Bitcoin will continuously increase in it's value, and while it does, Fiat currencies will steadily decrease in value; therefore it will provide an additional boost to the value of Bitcoin to grow even stronger.

Hardware wallets are the most secured, there is no doubt about that, however at first, it can be difficult to understand how cold wallets work.

Once you receive a cold wallet, at first you have to create your Pin code. Next, it will ask you if you are backing up an existing account, or creating a new wallet.

So choosing to build a new portfolio, it will generate 24 worlds that you have to write down on a piece of paper which comes with the wallet.

This is also known as a seed recovery sheet. Once that complete, you have to update the firmware. That could take another five to ten minutes.

After that, you have to choose what type of cryptocurrency wallets you want to download on your wallet. Hardware wallets such as Trezor or Ledger Nano S can support five different crypto wallets at the same time.

Of course you can choose from more than five currencies, however, let me give you an example: Let's say that you want to have Bitcoin, Litecoin, Monero, Dash, Zcash, which is not a problem.

However, a week later you want to add another crypto wallet to your hardware wallet, for example, Ethereum, you will not be able to do so until you delete one of the existing crypto wallets.

So, there are pros and cons to it, however to setup a cold wallet you might require to allocate 30-40 minutes for these purposes. Installing a mobile wallet is less than a minute; but you have to respect the security that comes with a hardware wallet.

Additionally, if you lose your equipment or get it broken, you might choose to purchase another one, and having those 24 random words you have previously wrote down, you can re-install all your crypto wallets in no time.

A cold wallet is an absolute must to have, especially if you thinking of a long term investment in Crypto, like I do.

Even I keep on telling you that you must have a cold wallet, there are few other cons that you should be aware of, or you should prepare to.

First is this: Do not keep the recovery sheet at the same place where you keep the hardware. Why is that?

Imagine that you get a burglary, and they find a cryptocurrency hardware wallet, and next to it another piece of paper that says: Recovery sheet, and below the 24 words that you wrote down.

They don't even have to take your hardware wallet. All they need is a recovery map, as they can recover your wallet, using another device that is compatible with the one you have.

As you see, even it is the most secure wallet that exists; still, all can be lost by not taking extra measures.

Let's take another example. This time there is a fire in the building. Again it is good to have a recovery sheet, however, if that gets lost for any reason, well, if you still have the device, you have two choices. One is to look after that device forever, or send every fund to another device that is backed up, and you know exactly where that recovery sheet is.

You might choose to put it somewhere safe like in the cloud, so you can access it even in the future at any time, however having all those 24 words together not such a good idea, especially keeping them online.

So what you can do is this: Keep six words on Evernote, another six notes on Facebook, another six on Gmail, and another six on Yahoo storage. This is just an idea.

However, it's completely up to you but let me tell you another issue with too much security. I will assume that you are men, and you have a wife, who isn't really into the crypto world like you.

Most probably your wife has no clue what Bitcoin is, not to mention hardware wallets, or recovery sheets, and platform exchanges.

Let's imagine that you have 10 Bitcoin and each Bitcoin worth around 4000 dollars, but something happens to you. Let's say that you end up in the hospital in the coma.

I am sorry that I come up with the worse examples, but my point is this.

There is no sense to save all that crypto money for your son or wife, if they never going to be able to access it.

So, because you put the recovery sheet in 4 different safes, and each safe is in various banks, and each bank is in various country, you might choose to have so much security that not only hackers, but even your loved ones won't be able to access those funds ever.

What I am trying to explain is that you should teach your loved ones, or you should tell them that you have written down everything step-by-step, how to access those funds, in case of emergency.

Probably not everyone will do that. However, I think it's worth to mention to you, so you can decide how to go about it.

# Chapter 11 – Wallet recommendation

Some questions that you might want to ask yourself before choosing a wallet:

- *Do you need a wallet so you can use it all the time, every day?*
- *Do you need a wallet so you can keep on buying crypto currency and holding it for long a time?*
- *Do you need multiple currencies or only Bitcoin?*
- *Do you want anytime access, or only sometimes?*
- *Do you want to get paid in Bitcoin?*

Once you have settled and understood your need, you can choose what best fit your requirements, then go for the wallet suits you best. That's being said lets late a look at what I recommend that is a must have.

## Hot wallet recommendations

### Blockchain

I have explained already how Blockchain wallet works, and its limitations when it comes to cryptocurrencies. However, this is my number 1 recommendation to those, who are entirely new to Bitcoin and other cryptocurrencies. The platform can be reached on blockchain.info

# JAXX

Jaxx is one of the most popular mobile wallet, and to be honest, everyone's favourite.

Jaxx supports multicurrencies, such as Bitcoin, Ethereum, Ethereum Classic, Litecoin, DASH, Zcash, Monero. Jaxx supports 9 different platforms, that including Windows, Apple and Linux desktop, Android, IOS mobile and tablet, Google Chrome and Firefox extensions as well.
Jaxx has an excellent user interface that gives you comfortable easy to understand, easy to use experience.

Having on multiple devices, it synchronizes to another devices too, same as Blockchain wallet does. Jaxx appears to be slow sometimes, however this could be, because it is not open source, and it supports multiple currencies.
Jaxx can be found at: jaxx.io

# Cold wallet recommendations

## TREZOR

Trezor is a hardware wallet, that currently one of the best in terms of security. It is an excellent cold wallet, especially to store Bitcoin on it.

Once you have a Trezor, you will know why it is so secured as it has a screen, that only you can see. I mean, the Trezor screen is not visible on your computer, therefore, hackers can't get to it, and your private keys are completely off line.

The interface is very easy to use, and I would recommend it to anyone. It also have a web interface for easier use, however the screen is built in to provide additional security.

If you want to buy it from the source, you can buy one on following the link:

# https://trezor.io/

It's open source, therefore if case you lose it, you can buy another one, or similar device that supports the same functions, then back it up very quickly.

It only costs 89 Euros, however, when I bought it, I have also purchased a Cable for an Android phone, and paid for DHL shipping another 26 Euros

Then I got charged 21% VAT, so instead of 89 euro, somehow I had to pay 145.20 Euros in the end.

I am not saying they are scammers, they are not, but you will get charged for postage no matter where you order from, and the VAT itself was 25.2 Euros, which I wasn't happy about either.

Anyways, it is coming from the Chech Republic, and if you live there, you might not require paying the VAT, but I wouldn't count on that.

As I mentioned earlier, it is not a scam, and it is a must have, however before you consider buying one, make sure that you are aware of additional charges.

In case you don't believe me, I have taken a screenshot of my order, and as you can see, I have ordered a White color.

You have two options when it comes to the color choices: Black or white.

Another issue that was not in my favor is when I bought it, there was a warning message that it would take six weeks for delivery.

However, once you move on, and click on payments, you have options for DHL delivery that is between 3 to 5 days costing 26 euros, or traditional delivery that is 4-8 weeks for 12 Euros.

So I have chosen the DHL, as I wanted to store all my Bitcoin on a secure wallet as soon as possible.

Still, after six working days have passed, I have sent them an email asking them what is happening.

They replied that there is a warning message about the delivery that takes at least six weeks.

I was outraged, and my point was that if it's six weeks as a minimum to deliver, then why do they still offer 3 to 5 days DHL delivery.

In the end I got it after five weeks; however it is a long waiting time, but again it is a must have, so it's all right.

I have survived all the waiting, and my Blockchain wallet didn't get hacked while I did it.

Still, I would recommend to buy a cold storage wallet first, only then start buying Bitcoin, especially in large quantities.

## Ledger Nano S

Ledger Nano was no different when it comes to delivery, I had to wait 5 weeks, and few more days too.

The difference was that the Ledger Nano S was only 69.60 Euro and it's already including TAX. Of course, I had to pay for the postage another 16,28 Euros, this time for UPS. Totalling the full payment of 85,88 Euros. To buy it from the source, just follow the link:

**https://www.ledgerwallet.com/r/e101**

Ledger Nano had another trick, and that is this: Once I made the payment, I got a confirmation e-mail, stating this:

*Due to a very high level of demand, and in accordance to the delivery terms you accepted at ordering time by checking the confirmation box, your order will not be shipped before September 4, 2017 and then will be delivered to your address 2-5 days later.*

Now bare in mind this was back in June 24th of 2017, meaning I should wait 11 weeks for delivery. I was really sad, and for some reason I can not recall any checkbox when I first ordered one.

However, since I have ordered another one for my father, this time I got them checkboxes for sure.

Anyhow, as I mentioned I only had to wait less then 6 weeks for it's arrival, and I am vary happy with it.

In fact, I prefer Ledger Nano S, than Trezor. Ledger Nano is much more beautiful, at least to me, and it's screen is more likeable.

It's also cheaper, and the knowledge is pretty much the same as the Trezor. When it comes to the build quality, also the Ledger Nano is what I would recommend.

The ledger Nano S has a capability to take a weight of a car, so my first choice is Ledger Nano S

My recommendation is to use a cold wallet, so you can not be hacked, and again, as I mentioned before, I would recommend to go for Ledger Nano S.

## https://www.ledgerwallet.com/r/e101

Not only because it is cheaper, but it is not that scammy like Trezor, where you have additional VAT fees and super expensive, also the Ledger Nano looks much more better, but in the same time I have them both Trezor as well Ledger Nano S.

# Integration using Ledger Nano S

Ledger Nano S also compatible with another nine different Cryptocurrency software wallets, that can integrate as below:

- Ledger Wallet Bitcoin
- Ledger Wallet Ethereum
- Ledger Wallet Ripple
- Copay
- Electrum
- Mycelium
- MyEtherWallet
- GreenBits
- BitGo

This is the best wallet I can recommend, however I wanted you to know that are other options too. Once again, you should decide it for your own reasons first, then you will be able to make the best choice for yourself. Again, there are many wallets that you can use for free right away, and you probably should as your first wallet. However, when it comes to investing, hot wallets are not safe, therefore I would only recommend a low amount like 100 dollars as your first investment.

Anything less then 500 dollars are not that attractive for Hackers, however it doesn't mean that you are safe, therefore only invest in the amount of Bitcoin as much as you are ready to lose, in case you get hacked.

In the meanwhile, you can buy a Ledger Nano S, as it will take few weeks to arrive, and while you are waiting, you can read more books on Blockchain and Bitcoin for better understanding, as well to become more comfortable with the technology.

Once you receive your Ledger Nano S, you can start to invest larger amounts, and keep it safe on your Ledger Nano S.

### https://www.ledgerwallet.com/r/e101

I have both Trezor as well Ledger Nano, and I also know that some people prefer Trezor over the Ledger, however in my opinion, Ledger Nano S is the best.

Trezor was released in 2013 and it's design doesn't look so appealing, however Ledger Nano S was released in 2016, with a lot better looking hardware, or at least to me, it is more stylish then the Trezor.

I believe that Ledger Nano S has a way better looking then Trezor, however if you think otherwise, it's fine, but please bare in mind that Trezor is more expensive.

In order to buy them again use the following links:

**Trezor:**
https://trezor.io/

**Ledger Nano S:**
https://www.ledgerwallet.com/r/e101

As I mentioned, I had to wait for both of them, however if you do not want to wait for weeks, you can check on eBay or Amazon where others might sell it too.

I have realized that second hand devices are much more expensive. In the same time, it might worth it.

What I mean is first I have tried to buy it on Amazon, even it was more expensieve by a third party seller, still I was going to buy it from Amazon, as I trust Amazon delivery, their customer service as well it's fast.

I have Amazon Prime too, and pretty much anything I order from Amazon, I do get it as a next day delivery.

So the problem was when I wanted to buy the Ledger Nano S is simply was the availability. Literally Amazon was out of stock too, so, even I was going to pay more for the Ledger, next day delivery would have been awesome, but no stock.

If you are in the rush, or just want to use Hardware wallet right away, like I wanted to, I would advise to check Amazon first, as you may get it lot more faster if they have in stock of course.

## Chapter 12 – Bitcoin ATM-s

That's right, there are machines called Bitcoin ATM-s. They are connected to the internet, as well looking very similar to a traditional ATM-s.

However the purpose of these are to convert Bitcoin to Fiat currency or vice versa. In 2013 the first Bitcoin ATM has been installed in Canada, and it has been a continuous increase of Bitcoin ATM-s ever since. Currently, as of 2017 August, there are 1493 Bitcoin ATM-s around the world, operating in 57 different Countries.

## Types of Bitcoin ATM-s

It worth to mention that are many different types of Bitcoin ATM-s, and some are only operate in one way, however, some has a two-way function. It also depends where you use the Bitcoin ATM-s , as if you want to convert your existing Bitcoin to cash, you will probably receive the local currency of that country.

Other issues that you might encounter is the fees. There are Bitcoin ATM-s that are operating with no fees, however some others can take as much as 5-10% fees once used.

The legacy of Bitcoin is that are no fees, however, there are many operators who paid for producing such machines, as well costs to pay the electricity bills, rental fees, therefore some might charge for a certain fee.

It is advisable to check the fees before using one, however the comfort that it provides are extremely helpful.

## How to use Bitcoin ATM-s

In case you wonder how it works, I can tell you from experience that is relatively easy.

Let's assume that you want to buy some Bitcoin, using traditional Cash such as dollar. The checklist that you should have is this:

- Smart phone with internet connectivity: Any types of smart phones are ok.
- Hot wallet downloaded on the smart phone: Blockchain wallet or Jaxx
- Dollar bill: Ten, twenty or any dollar bill that you want to convert into Bitcoin.
- Bitcoin ATM near you: You can find a local Bitcoin ATM near you by visiting this link:

## https://coinatmradar.com/

Once you have downloaded one of the wallets I have recommended, or any other Cryptocurrency wallet to your smart phone, visit a local Bitcoin ATM.

In case you think that Bitcoin ATM-s are placed in some dark hidden street, let me tell you that normally the Bitcoin ATM-s are in a public place such as restaurants, pubs, or local shops, places that are many people visit daily.

**Buy Bitcoin**
Once you are there, and happy with the fees that the ATM will operate, do the following:

Step 1. Click on the machine's screen: Buy Bitcoin

Step 2. Open your Hot wallet on your smart phone, and select: receive. This will bring up your QR code on your smart phone screen.

Step 3. Hold your phone to the Bitcoin ATM-s screen, and let it scan your QR Code.
In the meanwhile the ATM will tell you to insert a bill.

Step 4. Feed the paper bill to the machine. This time you can feed as much as Bitcoin is available, however if you do it first time, you might try it only with a 10 dollar bill.

Step 5. Click send Bitcoin on the ATM. The Bitcoin ATM will perform the transaction, and you might see on the screen something like: Sending Bitcoin. It would take around less then a second.

Step 6. Check your smart phone for notifications: You should have on your screen something like: New payment received. Of course it depends on what wallet you are using. Once you happy with the Bitcoin that you have received, you may also try out how to sell your Bitcoin for cash. Again, you should check beforehand, making sure that you visit a two-way function Bitcoin ATM, but of course if you only want to buy Bitcoin, than you can visit a one way operating Bitcoin ATM too.

This case let's assume that you are ready to convert your Bitcoin to cash, using a Bitcoin ATM. Please note that some of the Bitcon ATM-s are the minimum limit is at least 5 dollars, meaning if you only want to sell Bitcoin that worth only one dollar, it might not be possible. It is better to check before, however normally the minimum cash to take out, is at least 5 dollars.

**Sell Bitcoin**

Step 1. On the Bitcoin ATM, select: I want Cash. This will bring up the next screen, listing 5, 20, 50 that you can tap on. Note here: some ATM-s have the same functions like traditional ATM-s where there is an option for: Select other amounts, however some Bitcoin ATM-s have no numbers to create your special amounts, instead, you can keep taping on the listed amounts. For example if you want 30 dollars, but there is only option for 5, 10 and 20, you can tap on the 10 three times. This will provide you with a 30 dollars.

Step 2. Choose the amount you want: For example, select 5, for five dollars worth of Bitcoin.

Step 3. Tap on the next screen: Cash out. Once you tap on: Cash out, the ATM will generate a QR code. This will be for you to scan it, using your smart phone.

Step 4. On your smart phone, open the hot wallet and select send.

Step 5. Hold your smart phone to the ATM and scan the QR code that the ATM has generated previously.

Once you have scanned the QR code, using your smart phone, it will generate the transaction that you will have to confirm in the next step.

This step, your phone will calculate the amount of Bitcoin that you have to send to the Bitcoin ATM's address.

Step 6. On your smart phone, select: SEND. Once you tap on SEND on your smart phone, it will ask you to confirm it again by opening another window.
Here, you will see how much Bitcoin you will send to the address of the ATM.

Step 7. Select: Confirm. This will send the transaction to the ATM. This will take another second or two. However once complete, the Bitcoin ATM will come up with a screen: Bitcoin received! > This screen will quickly change to another display where it will say: Dispensing... Please take your cash.

Step 8. Take your cash! Check on the Bitcoin ATM below for your cash, and take it. There are many different Bitcoin ATM-s and some functions different then others, however the process is somewhat the similar.

I have not try them all, however I did try out two of them already and they both were working just fine. I have tried one that was charging me for 3% and another Bitcoin ATM that had no fee.

In the end of the day, for the convenience of selling and buying Bitcoin instantly, and anonymously it's awesome.

In case you are afraid of being hacked online, this is one of the safest way to buy Bitcoin. Again, I am not sure where you live, however using coin atm radar, you should be able to find one close by to your home or work.

https://coinatmradar.com/

# Chapter 13 – Best Online Trading platforms

Back in the day you were lucky if you were able to find at least one Bitcoin trading platform. However, nowadays, there are so many that you cant even count them. The real problem is not to find one, instead making sure that you are not getting scammed on some fake cryptocurrency trading platform.

I will explain with more details on how to recognize fake websites and scammers on the next chapter, however first I would like to introduce some great platforms that you can use in the future.

Please note, this book has been written in August 2017, and I am pointing out the best online trading platforms that are exist today.

Instead of analysing each online trading platform, the question you should ask is this: how to find genuine online trading platforms without getting scammed.

In order to find platforms, as a starting point, you should be looking at cryptocurrencies that are on the market since years such as Bitcoin. Bitcoin is not a scam, in fact Bitcoin is the strongest cryptocurrency of all, and of course the first even cryptocurrency that was created. Bitcoin exists since 2009 January, meaning more then 8 years already on the market and stronger than ever.

Of course you already know that by now, however just to point it out again, if you follow Bitcoin and it's markets, by what platforms it is mainly traded, you should be able to find those online trading platforms that are completely legit. Where to start?

The platform that you are looking for is what I already introduced previously that is called:

## coinmarketcap.com

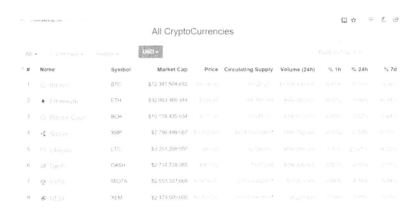

https://coinmarketcap.com/all/views/all/

Once you have navigated to coinmarketcap, you will find all cryptocurrencies that are on the

market. Next if you click on Bitcoin, it will take you to the next page, where you can find more details about the currency:

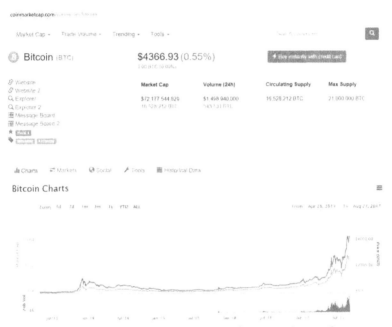

In this page there many information that you can find about Bitcoin, including:

- All the websites where you can find out more information.
- Market capitalizations,
- Additional tools,
- Historical Data, and so on…

It's easy to just get lost here, as there are some great information about Bitcoin, however, our main focus is to find a genuine online trading platform, therefore you should carry on by selecting the menu called: Markets:

By selecting the menu option: Markets, the new window will open where you can find information about:

- All Bitcoin Markets,
- Source – these are the platforms we are looking for.
- Pair – meaning what Bitcoin can be exchanged into.
- Price – This is the price of the Bitcoin on each of those markets

- Volume – This number represents the percentage of all Bitcoin that is currently on the market; however the number is unique to each platform.

You can also find on the right another button that now shows USD, however, if you click on it, you should see all other currencies that you can trade with when trading Bitcoin.

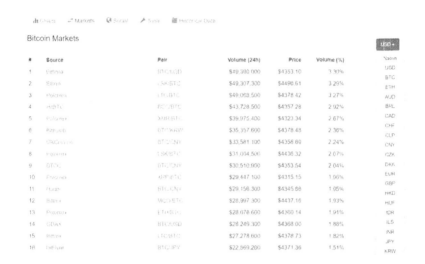

At this moment Bitfinex has the most Bitcoin on the market, and it's pairing BTC/USD, meaning you can buy Bitcoin to US dollar or vice versa.

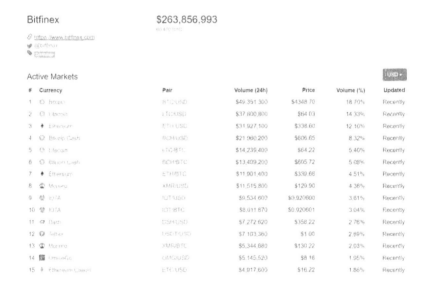

Next by clicking on Bitfinex, it will take you over to the next page, where you can see all the currencies that this site is currently trading with, as well the website on the top right corner, and its Twitter account:

By clicking on their twitter account, you can see how much engagement the site has. Twitter has all the complaints as well appreciation for the platform, therefore by taking a quick look for educational purposes, is a good idea.

For example, you can see that are more than 56.8K followers, and that can tell you lots too. If it was a new trading platform, you would probably see lot fewer members, and that case would be advisable not to get involved yet;

however more than 50K followers are just fine. Also you can contact others and ask their opinion about the platform.

Of course this is a genuine platform; therefore you can also visit their website at the link provided on coinmarketcap:

https://www.bitfinex.com/

Once you have reached the platform, you can just register for free, and start trading Bitcoin or other cryptocurrencies. Bitfinex is one the most respected Online Trading platform for multiple reasons.

Unfortunately, Bitfinex was hacked in 2016 August, and some of the traders also have lost their value, as some of them have left their cryptocurrencies on the platform instead saving them to a hardware wallet.

The hackers have taken 120K Bitcoin at the time, and even it seemed that Bitfinex would go into administration like some other crypto traders previously, Bitfinex has recovered.

In fact not only recovered, but paid in full to all their traders who have lost their value when the hack has happened.

Even it has taken eight months for them to reimburse all their investors, they could have just closed their doors, however instead they have looked after their investors, and paid back all their losses in full in 2017 April.

Since that event, there have been even more investors than ever, as they have provided an example of trust and long term relationship, when it comes to investing in cryptocurrencies.

I wanted to show you an example on how to find an online trading platform by only following links, however, as you can see there are many platforms to choose from.

Because of that fact, I will cover some other platforms that I have been using previously.

First things first, even Bitfinex have paid back all their investors, which is nice. Still, I hope you understood by now, that no online trading platform is safe. They all can be hacked, and probably will be hacked some day.

Because of the way that cryptocurrency works, there is very little that police can do, if they might take away all your Bitcoin.

Simply, there is no guarantee that once these platforms are hacked, they will give you back all your losses.

## Coinbase

Coinbase is one of the most convenient exchanges out there to invest in, or withdraw money if you live in the US or other supported countries. The website works just perfectly, and you can reach it by following this link:

### https://www.coinbase.com/?locale=en

The interface is user-friendly, and it is recommended to anyone, even if you have no technical background, you will be able to navigate the site quickly.

Additionally, they have some excellent charts to see when it comes to technical analysis. If you have a problem, you can log a support ticket, and the responses are usually speedy.

Coinbase is also a good place for beginners, however when it comes to authentication, you should make sure that you choose to have a google authenticator, instead of getting verified by MMS.

Using MMS to authenticate is one of the ways that your account can be hacked. Therefore it is highly recommended to take the right security measurements.

**Poloniex**

The site can be reached by clicking on the link:

### https://poloniex.com/

Poloniex has countless of crypto currencies; however the highest volume of exchanges are Bitcoin and Ethereum. This site does not except

US dollars; instead you must use Tether, that is another crypto currency which has a purpose of a US dollar. Meaning 1 tether is always around 1 dollar. On Tether, you can find more information on their website:

## https://tether.to/

Tether is a digital value of the dollar. However, some online platforms such as Poloniex do not want to trade with the traditional dollar, instead tether.

Back to Poloniex, they have become so overwhelmed due to too many investors; they are literally super busy for the next few years probably.

What you have to understand is that, when you register on Poloniex, they will check your account, and eventually verify it before they allow you to trade on their platform. Now that's all ok; however, they have so many new customers recently, that some people have been waiting three months to get verified.

However, some others don't even get replied after three months. The customer service became horrible over the last six months, and many people have just moved on and started trading on other platforms instead.

## Kraken

Kraken can be used for those living in the European Union countries. Reaching Kraken, only use the link:

https://www.kraken.com/

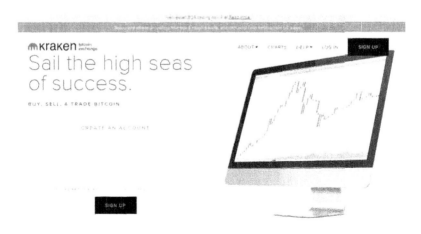

Kraken accepts Euro as well British pounds; therefore it's one of the best choices to Europeans to trade on.

Kraken uses a Tier system, meaning you can start trading as a Tier 1, where you can exchange between all currencies, but account funding is limited to digital currencies only.

Tier 2, is where your daily limit is 2000 dollars worth by both, depositing, or withdrawing, and in both, Fiat currencies as well in cryptocurrencies.

Tier 3, it allows the same functions as Tier 2, however this time the daily limit is 25000 dollars. Also worth to mention that if you want to withdraw Cryptocurrency, your daily limit is 50000 dollars.

Same as with other crypto trading platforms, you must get authorized, to get started on Kraken, which has taken me a week; however I have heard that some people have to wait longer than that.

The interface is very friendly, and there are some excellent charts too; however the security is that I want to point out again. Two-factor authentication is what you want to use. Meaning, you will use your password as well your Google authentication using google authenticator. This is what you have to set up, as the system will not do it for you.

You must have a smart phone, and you must download google authenticator, is a free application, then set it up for better security.

I have covered all those online trading sites I have used previously, and those are all legeit genuine websites.

There are many more online trading platforms that are also legit; however, I have not tried them out yet; therefore I can not review them.

All through, you may find other websites through coinmarketcap.com, and they all should be legit.

Still, you should do some more research and make sure that you are comfortable using any cryptocurrency trading site.

## Chapter 14 - Be aware of scammers!

As I mentioned previously, there is a huge industry around Bitcoin and other cryptocurrencies. Still, many people just beginning of learning about Bitcoin, and most of them who does, want to buy some. Unfortunately, I have to tell you that hackers are one of the worse enemies when it comes to Bitcoin.

Hackers can hit anytime, at anyone, and they can steal Bitcoin from personal accounts, as well from large online trading companies.

Even experienced people do get hacked time to time, and the reality is that most people are only taking security seriously once they got hacked.

It's nice to own some Bitcoin for weeks, then months, and keep on buying more and more, however, if you do not take extra security measurements, one day you might wake up, for your wallet being emptied out.

Any purchase you have, you must make sure that you use a hardware wallet and save all your values on it. Still, my number one recommendation is Ledger Nano S:

## https://www.ledgerwallet.com/r/e101

I can not emphasize enough, that authorities, such as police or even your bank, will not be able to help you once you get hacked.

There are some trading platforms who might get hacked and will repay you in full, such as Bitfinex did. However, there is no guarantee that they will do it again, in case history will repeat itself.

Now that's being said, as I mentioned there are many new comers to this industry, and beginners who are often having no knowledge, they hear that the value of Bitcoin has been increased dramatically; therefore they quickly want to get some.

Anytime when you or anyone you know is so excited, and want to rush to the market to buy some Bitcoin, please do not do it, and those you know, tell them that is one of the worse things they can do.

Sure thing, you hear that Bitcoin is a good investment, and you have money to invest, so let's do it right? WRONG!

There are so many scammers around the web, that is very difficult to say which ones are legit, especially for those, that are novice to this market.

Once you have the experience, you will know how people with bad intentions are trying to scam people, using Ponzi schemes and other methods. How do they reach potential victims?

Well, they go where the people are. Those might be Facebook, Twitter, or youtube, and begin advertise themselves.

Trying to make you believe, they will give you a huge profit margin or even make you a millionaire. They are using techniques such as creating a fake facebook account, or youtube channel, as well counterfeit websites.

First of all, anytime you hear things like double your money or making a profit every day, even if Bitcoin market is down, there is something dodgy going on.

Unfortunately, there are many people losing thousands of dollars because they fall for some of these scammers or thieves, and there is not much they can do about it.

Let's begin with some examples as facebooks, and twitter comments.

## Scam No 1.

*To increase you Bitcoin investments, you can start trading today with only 10 Bitcoin, and earn up to 30 Bitcoin in less than a month...*

There is no point reading this any further, however, let's get the first sentence right.

So if I invest <u>only 10 Bitcoin today</u>, that is currently around 45000 dollars, <u>I can earn 30 Bitcoin in less than 30 days</u>.

So basically in 30 days, I could make 135000 dollars right? Wow!

That's sounds good! In fact, it sounds too good to be true! That would come to 4500 dollars daily profit.

This is obviously a scam. I am not even going to ask how much he or she is making in this business model, simply because there is no point, so look at another example.

## Scam No 2.

*Anyone need Bitcoin is very cheap rate so please Whatsapp me my number...*

Right! So anything like this, you ever see, <u>do not send a text message</u> at all.
There is an enormous hacking industry going around stealing people's phone numbers, then trying to use your number to their advantage.

## Scam No 3.

*Cloud mining – click on the link, free start, payout after ten days.*

OK, so there are many different kinds of cloud mining; however even some are real such as Genesis mining or Minergate; still, I wouldn't click on links that I am unsure of it.

I am using Minergate, and my friend is using Genesis mining; however, I can not recommend others as I don't know any other legit company.

However, if you want to mine some Bitcoin or another cryptocurrency, you can use Minergate for free. ( more on this in the bonus chapter )

**https://minergate.com/a/f5dccb84d2696 b16a1c8bced**

Another famous mining facility is called Genesis Mining, however this is not free to use, and it is not you but them who mining for you:

**https://www.genesis-mining.com/**

## Scam No 4

Onecoin scam. Onecoin has been introduced as a new cryptocurrency; however, it has turned out to be the biggest scam in the crypto world.

Unfortunately, many people have lost their life savings, due to investing into this fake currency. Those individuals who have been promoting this currency, have been sent to jail already in India as well in Dubai.

The company has formed in Bulgaria with the intention of hoping to find gullible people around the world and sell them, making them believe that one day, they all become rich.

The problem is that you can only buy Onecoin from the distributors, however once you are ready to sell it, you will not be able to do so.

They will not repurchase it from you, and if you think that you can find an online exchange, you are wrong.

No one dealing with this, therefore it has no value. As I mentioned before coinmarketcap.com is where you can find all crypto currencies, however, you will not be able to find Onecoin there.

How do people fall for Onecoin? Onecoin was promoted as a possible next Bitcoin, also seen adverts where they have advertised Onecoin and Bitcoin in the same sentence.

This, of course, cought some people's attention, and begin to purchase it.

However, once they have realized that it has never made it any further, people wanted to sell it, but it was too late.

Please avoid investing in any new currency that you can not find on coinmarketcap.com.

## Scam No 5

<u>New-age-bank scam.</u> This website has been shut down for a while now, however at the beginning of this year was active, and many people have fallen for their scam system.

What they have been promoting, is that if you invest 0.05 Bitcoin, they will return you 0.1 Bitcoin in 30 days, however, if you spend 1 Bitcoin, they will return 2 Bitcoin in 28 Days.

Of course, if you invest 5 Bitcoin, they would return 10 Bitcoin in 24 days. The VIP package is where you would spend 10 Bitcoin, and you would get returned 20 Bitcoin but this time in 23 days.

So, basically, thay will double your money either way, however more Bitconi you send them, faster the doubling it will be.

Now, this is sound too good to be true, so please anything like that, you must avoid.

All through their websites are now shut down. Still, there were hundreds of people who claimed that were scammed, while they have been operated.

The reality is that I could go on and on about these sites, so let's look at some common issues they all have.

**Scam alert No1:**

They put pictures of famous people on the website. However, you can not find out exactly who is the CEO, or the link to About Us does not work, or worse there isn't one.

There are some occasions where the Contact Us menu option isn't working or once again there isn't one on the site.

**Scam alert No2:**

Unreasonable high return. Anyone who claims that can double or triple your money is 99% scam. In the crypto world, no one can tell you exactly what the market will bring us.

Of course, it is possible to double or even triple your money, however, if someone guarantees that for you, that is possibly a scam.

The value of Bitcoin has increased since January 2017 from 900 dollars to 4500 dollars by the end of August 2017.

It is true. However there is no guarantee for that to happen in the next year, even I really would like that to happen, it might not going to.

## Scam alert No3:

Unsure the purpose of the company. Sometimes if you can click on the About Us menu option (if it works), and you find that you are not 100% clear what the company does exactly, that could mean a possible scam.

Some might say, that they are trading for you, but then how do they make up for their profit? Well if you can contact the company, you should ask them questions like:

- What is the company does exactly?
- How do they guarantee my investment?
- How do they ensure security?

If you get a response that is a bit wishy-washy, well then probably is a scam. If they don't even reply to you, then again, it is possibly a scam.

## Scam alert No4:

The particular coin is not listed on coinmarketplace.com. If you can not find a coin that you are interested in coinmarketcap, then you should probably turn around and run away from that coin, as its probably another scam.

If there is a new cryptocurrency, do not go to their own website and start investing. First you should always check coinmarketcap.com, simple as that.

**Scam Alert No5:**

The coin is centralized. I have explained this before on Volume 1 Blockchain book, that once a system is centralized, it is controlled by an individual or a company.

If it has a central Server, that can be monitored; there is no guaranty that they will not go and alter it or to make any changes.

Again, I could go on and on about possible scams, as well how to recognize them, however, if you are a newbie, you might find it useful that, there is a dedicated website for Bitcoin scammers, called:

### http://behindmlm.com/

This site is a collection of possible scams, where also people reply to one another, explaining experiences, of different websites.

Again, please be extra vigilant, and research on any trading company before you would invest into it. Double check their site and make sure everything works, as well checks out ok.

Also if you can contact them, then by all means do so, and ask questions that might concern you.

Genuine company should get back to you within 24 hours due to time differences, and they would provide you with an answer that you were looking for.

## Chapter 15 – Bitcoin Trading

I have explained previously how to find genuine cryptocurrency trading platforms, however I have never got on the topic of trading with Bitcoin.

The reality is that I am not a Bitcoin trader, all through I do make some moves here and there, still, my belief on trading is not something that is beneficial for long term business, as it comes with lots of stress as well a huge risk.

I do believe in Bockchain, and Bitcoin too, as a technology. Once you have enough knowledge of how the technology works, and what issues it has, that require fixing, you can predict changes in it's value.

Of course there are other occasions too, besides it's technology features, when it comes to its value and that is fame.

For example, this year the value of Bitcoin was 1700 dollars, however when Wannacry ransomware has hit worldwide, by an anonymous hacker group, they have asked for ransom to be paid in Bitcoin.

Within a week, the price of Bitcoin has gone up to 2400 dollars. Ransomware has hit more then 150 countries, by disabling the logon functions on more then 200,000 computers worldwide. It is important to note on that incident that Wannacry has not reached to individuals, instead Hospitals and government agencies, as well Police stations. Some speculate that many organizations have invested heavily into Bitcoin, in case another attack happens, so they could pay up the ransom right away.

All through it makes sense, others including me, also believe that is a side effect of Bitcoin made it to the news once again. It has been learned in previous years, that each time Bitcoin was in the news, there is an enormous amount of people who began to take interests about it's nature.

Those they do so, usually take it a bit further and realize the potentials, than start to invest into Bitcoin. What also have been learned that each time when the market capitalization grows, parallel the value of Bitcoin grows too.

Having said that, if you can follow the news of Bitcoin, your odds of trading with Bitcoin, are certainly will increase. I have researched a historical data for Bitcoin, and I have compared the market capitalization to the value of Bitcoin.

The date is from coinmarketcap.com, and if you click on this link, it should take you there too:

**https://coinmarketcap.com/currencies/bitcoin/historical-data/?start=20130428&end=20170828**

In case you don't feel like going through years of data, I have taken some notes for easier understanding what has happened over the years.

Please note that coinmarketcap only generated a report since 2013; therefore I wasn't able to get earlier figures.

However, I believe that only looking at the data of 2017, can be useful already. I have followed the market capitalization, followed by the date and the value of Bitcoin, upwards until recent dates.

| Market Cap | 1 BTC in US $ | Date |
|---|---|---|
| 1,500,520,000 | 135.98 | Apr 28, 2013 |
| 870,912,000 | 80 | Jul 08, 2013 |
| 1,547,640,000 | 140.89 | Aug 31, 2013 |
| 3,129,190,000 | 304.17 | Nov 07, 2013 |
| 5,029,340,000 | 437.89 | Nov 15, 2013 |
| 11,124,900,000 | 1001.96 | Nov 27, 2013 |
| 5,583,670,000 | 443.37 | Apr 10, 2014 |
| 2,853,930,000 | 211.73 | Jan 17, 2015 |
| 10,511,900,000 | 716 | Jun 13, 2016 |
| 15,667,900,000 | 989.11 | Feb 01, 2017 |
| 20,253,700,000 | 1280.31 | Mar 03, 2017 |
| 25,133,100,000 | 1618.03 | May 05, 2017 |
| 30,999,000,000 | 2004.52 | May 19, 2017 |
| 40,817,100,000 | 2581.91 | Jun 03, 2017 |
| 46,276,200,000 | 2899.33 | Aug 04, 2017 |
| 47,778,200,000 | 3290.01 | Aug 05, 2017 |
| 53,720,900,000 | 3293.29 | Aug 06, 2017 |
| 60,242,100,000 | 3949.92 | Aug 12, 2017 |
| 71,425,500,000 | 4455.97 | Aug 15, 2017 |
| 71,809,200,000 | 4416.59 | Aug 27, 2017 |

As you can see, it has taken years to reach 10 billion US dollars in market capitalization, however alone this year, since January, Bitcoin has reached 70 Billion in market capitalization alone.

Not to mention the other hundreds of Cryptocurrencies, totalling 159 Billion. Meaning

Bitcoin has only 45.1% dominance of the whole market.

Why am I telling you this? Well, because the value of Bitcoin has reached an amount that we have never experienced with any other currency, there is a tremendous volatility that comes with it.

Meaning, there are days when the value of Bitcoin goes up and down in hundreds of dollars, but for better understanding, let me show you an example of today.

I have taken a screenshot of today's Bitcoin charts, and I can see that today morning at 6am the value of Bitcoin has gone down to 4224 dollars, however later on at 5:44 pm the value has reached 4385 dollars. That's 161 dollars difference.

Please bare in mind that crypto market is a free market, open 24/7. Meaning you can sell, buy, exchange anytime you wish, from anywhere in the world.

There is no need for waiting until 9 am for the traders to open, neither the banks until Monday, in case of Saturday night you wish to sell some Bitcoin.

The 161 dollars difference over 1 Bitcoin it's a huge difference. However, if you think that is practically unfeasible to trade with, then you might wait for another day or two, and make a sale than.

Many people have left the traditional trading using fiat currencies and precious metals, so they learn and invest learning the crypto world instead. If you do have 10K worth of Bitcoin, and know your way around the crypto market, you can make 100-500 dollars a day, only trading with Bitcoin. Believe me, many people's new full-time job is to trade online, using Bitcoin as well other cryptocurrencies.

I am providing some facts on historical data, however, I have to mention that I am not a financial advisor, and please don't take these as financial advice.

Also, make sure that you understand, that if you choose to invest into Bitcoin, or other cryptocurrencies, this is a very high-risk market, and you should not invest more then what you are willing to lose.

As I mentioned, the crypto market is very volatile, therefore if you want to utilize the market volatility, you probably thinking like a trader. Investors, on the other hand, buy Bitcoin and hold on to it for weeks, months, even years to come.

What you can do, is this: You can become a mixture of both, an investor and trader blended together.

What I mean is that, instead of holding onto all your Bitcoin, and get angry at your self why didn't you sell it and bought more, try to take another approach. Lets imagine that you have

200 dollars worth of Bitcoin. What you can do is hold on to 100 dollars worth, and do not touch it whatsoever, even the market goes up or down.

This makes you an investor. The other 100 dollars worth of Bitcoin, sell it when you think the market is right for it, then wait until the market goes down again, and buy it back for cheaper.

This is the risky part of course. For example if you look at the historical data I have collected, you can see that the value of Bitcoin has gone deep down as 211 dollars in January 2015.

This event has made people believe that the time of Bitcoin has reached the end.

Even many people, who have bought Bitcoin previously for as much as 1000 dollars, they have sold their Bitcoin as they thought that Bitcoin will sink completely.

However, those who kept hold of it for years to come, has become wealthy.

## Bonus Chapter - Bitcoin mining with laptop

Many people get confused on the topic of Bitcoin mining, so let me lay it down once again. Bitcoin mining is a process of turning computer power into Bitcoin.

It allows you to generate Bitcoin, without the need to actually buy them. In case you heard about mining already, you might also heard that Bitcoin mining is impossible, using your home computer.

This is somewhat true however, there are other ways to go about mining with your laptop, or even with your Android mobile phone. Mining Bitcoin back in the day was relatively simple.

All you needed is a laptop or a desktop PC, and if you have mined Bitcoin, it might taken a day or two to generate one full Bitcoin. People used their home computer by utilizing its CPU power.

This is also known as CPU mining. However, over the time the hashing rate has been increased, therefore CPU mining has become impossible.

Next, people began to use their gaming PC-s, using graphical cards instead of their CPU-s, and that was somewhat profitable for a while.

This method is called GPU mining, however, as I just mentioned GPU mining became very difficult too.

The Next level of Bitcoin mining was done by ASIC mining machines. ASIC stands for Application-Specific Integrated Circuit.

These machines have been built for one purpose only, and that is Bitcoin mining.

Nowadays the hashing power has been so difficult, that even with the latest and greatest

ASIC mining machines are very difficult to mine Bitcoin. Imagine that having an ASIC mining rig that costs you around 2500 dollars, you could probably be able to mine a whole Bitcoin in every 2-3 months.

However, this is an average figure, as you might not be able to mine your first Bitcoin just in 5-6 months time.

Due to this reason, solo miners have began to create mining pools, were all single miners are able to participate their mining power.

Together are always able to mine Bitcoin, and they can also generate some additional Bitcoin by the transaction fees.

This could generate them on an average of 100 to 200 dollars a week.

Considering that these latest and greatest ASIC mining rigs will be exhausted within 2-3 years, it's a better profit then becoming a solo miner, and wait for your luck.

In case you are more like a miner type then an investor, and you don't want to invest either, in Bitcoin or expensive ASIC mining machines, I have a nice surprise for you.

Having an old laptop might be just enough for you to start generating your Bitcoin.

What you have to understand, is that any computer can mine Bitcoin, even it's old, and the CPU is not up to date, it will do the Job.

You will need to have a computer, a software called Minergate, and of course an internet access.

Minergate is available to be accessed on the following link:

https://minergate.com/a/f5dccb84d2696b16a1c8bced

Before you download the Minergate software, let me explain that using this software can be very demanding regarding using your computers CPU or GPU.

However, the good news is that you can set it up in the way, that it would never use more than 10-20 % of your CPU.

Once you ready to download the software, just click on the option called downloads.

Here you can choose what operating system you have on your laptop or computer, then just go ahead and download that software.

To download the app to your smart phone, you have to go to Google Play Store, and search for the App called Minergate.

After a successful download, you can start to mine many different kinds of cryptocurrencies, and the beauty of this is that you can convert them all into Bitcoin.

Of course, it is optional, as you might want to keep the coins for yourself, and turn them into Bitcoin a later day, but it's completely up to you.

Minergate supports the following currencies:

- Z cash
- Ethereum
- Ethereum Classic
- Bitcoin
- Litecoin
- Bytecoin
- Monero
- FantomCoin
- QuazarCoin
- DigitalNote
- MonetaVerde
- Dashcoin
- AEON
- Infinium-8

Supported currencies

Once you download the Minergate App and software to your computer, you have to register by providing your e-mail address, and password. Then, once you are ready to mine, you can only start by opening the software. Another great thing is that you can use multiple machines at the same time, and use them all as your miners. For example, I use two mobile phones, as well three laptops to mine mostly Monero. Monero, is a good choice, as its also supported on the mobile App. On your mobile phone, you can also make some adjustments, such as: do not mine when your battery is low, or only mine when it's connected to a WIFI.

I use both options, and it is great as once I am near to an access point, and my phone is authenticated to the local WIFI, the App starts up automatically. Once you open the software on your desktop, you can navigate to the menu options and click on view. Here you can tick or untick those currencies that you want to see or mine:

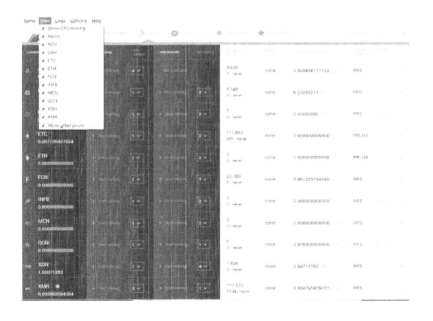

You can also click on smart mining, that will start automatically extract the coin that is the most profitable, which in most cases are Monero. By the way, Monero is currently worth 132 dollars and keep on rising.

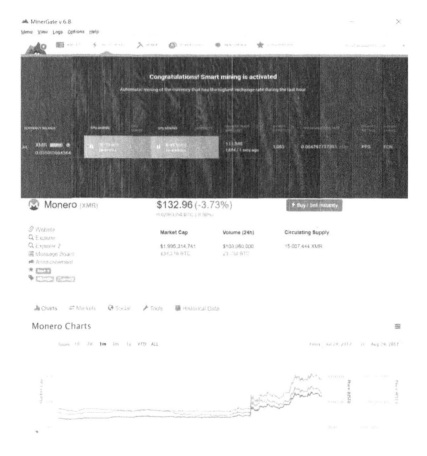

Additionally to the miner function and smart miners, you have free wallets to use for each of the cryptocurrency you mine. All the conis that you are capable of mining with Minergate, can be converted into Bitcoin, but as I mentioned, it is not necessary. Another function that Minergate has is the achievements section, where you receive different performance Prizes for the style of your mining.

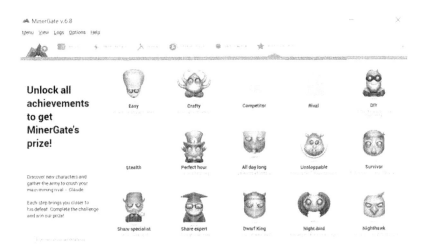

This is of course only for fun. However, it's a nice gesture from the creators. I will be honest with you, and let you know, that the smart miner all through picks up the most profitable cryptocurrency; however, it is not always true. The reality is that even other coins are not that lucrative.

Still, you might be able to mine more because everyone is mining Monero. I have realized that mining other currencies, you might get a better hashing power that would increase your mining capabilities, therefore not always Monero is the most successful, instead other coins, that fewer people are mining.

Again to start with Minergate, just follow this link:

## https://minergate.com/a/f5dccb84d2696b16a1c8bced

Being honest, Minergate will not make you rich; however an average laptop can generate you around 5 to 10 dollars a month, and if you can use more than one laptop like I do, as well mobile devices, such as phones, or tablets, you can generate more profit.

Another issue is the electricity bill. If you live somewhere, where the electricity bill is very cheap or even free, that's great; however, you have to understand that once you start mining, your computer will require more power. Therefore it will consume more electricity than usual.

Either way, I hope you will try it out and will like it too, as this is one of the fastest, and easiest way to start mining and generating your Bitcoin.

# Conclusion

Thank you for purchasing this book. I hope the content has provided some insights into what is really behind the curtains when it comes to the future of money.

I have tried to favour every reader by avoiding technical terms on how to invest in Bitcoin. However, as I mentioned few times, to fully understand how Bitcoin works, you may choose to read two of my other books on Blockchain as well on Bitcoin Blueprint.

*Volume 1 – Blockchain – Beginners Guide*
*Volume 2 – Blockchain – Advanced Guide*
*Volume 1 – Bitcoin Blueprint*

The Blockchain books are focusing on the underlying platform of Bitcoin, the technology called Blockchain.

Blockchain Volume 1, is for beginners; however, Volume 2 is very technical. Still, I did my best to use everyday English, and making sure that everyone can understand each of the technologies and their importance.

Bitcoin blueprint focuses on Bitcoin for beginners; however, I have fitted some interesting topics around the future of payments, using reputation systems with Bitcoin-Blockchain technology.

My upcoming book on Bitcoin, will provide more details on Bitcoin mining, profitability, and what exactly miners do.

How miners are often trying to manipulate the market, what techniques they use, and how they try to control mining pools, using super expensive ASIC mining hardware.

I will provide information and secrets on Chinese Bitcoin miners, as well their exclusively produced ASIC mining rigs, that no other miners capable of using.

Lastly, if you enjoyed the book, please take some time to share your thoughts by post a review. It would be highly appreciated!

# CRYPTOCURRENCY TRADING

*Strategies & Techniques for successful Portfolio Management*

## Book 3

*by*
## Keizer Söze

# Copyright

All rights reserved. No part of this book may be reproduced in any form or by any electronic, print or mechanical means, including information storage and retrieval systems, without permission in writing from the publisher.

Copyright © 2017 Keizer Söze

## Disclaimer

This Book is produced with the goal of providing information that is as accurate and reliable as possible.

Regardless, purchasing this Book can be seen as consent to the fact that both the publisher and the author of this book are in no way experts on the topics discussed within and that any recommendations or suggestions that are made herein are for entertainment purposes only.

Professionals should be consulted as needed before undertaking any of the action endorsed herein.

Under no circumstances will any legal responsibility or blame be held against the publisher for any reparation, damages, or monetary loss due to the information herein, either directly or indirectly.

This declaration is deemed fair and valid by both the American Bar Association and the Committee of Publishers Association and is legally binding throughout the United States.

The information in the following pages is broadly considered to be a truthful and accurate account of facts and as such any inattention, use or misuse of the information in question by the reader will render any resulting actions solely under their purview.

There are no scenarios in which the publisher or the original author of this work can be in any fashion deemed liable for any hardship or damages that may befall the reader or anyone else after undertaking information described herein.

Additionally, the information in the following pages is intended only for informational purposes and should thus be thought of as universal. As befitting its nature, it is presented without assurance regarding its prolonged validity or interim quality.

Trademarks that are mentioned are done without written consent and can in no way be considered an endorsement from the trademark holder.

## Introduction

Congratulations on purchasing this book and thank you for doing so.

Have you ever wondered to get into Cryptocurrency? Or perhaps you want to take your existing knowledge further, and begin to invest in crypto? In case you just wish to know more, and have a better understanding how thinks work in the Crypto world?

You might be thinking of trading with Cryptocurrency and increase your portfolio?

You might want to invest in the long term, and save funds for your retirement, or to your children, perhaps to your grandchildren? Either way, there is no better time than now, and I will explain why!

First, I will clarify what is Cryptocurrency, and it's fundamentals, so you can quickly grasp what Crypto is, and why you should take advantage of the market now!

Next, I will introduce some great tools, those that I personally use, and you will also need to get started.

Mainly, I will explain what is Wallet Technology, and teach you the differences between Hot & Cold Wallets.

I will explain in great details each of their Pros & Cons, then we take a closer look at the best Hardware wallets that currently on the market. In this section, you can take your pick: Ledger Nano S or Trezor, and learn where to buy them, as well how to use them!

If you prefer Cash only Business: I will show you how to locate Bitcoin ATM-s, anywhere on the world, and show you step by step how to buy or sell bitcoins for Cash!

After, I will provide you with the best Crypto Trading platforms, and you will find out which one best suits you, either you are from United States, Europe, or from the rest of the world.

Once you will have the fundamentals, I will provide you with same scamming techniques, often people fall for, and teach you how to avoid online scammers.

Next, I will introduce the term: pump and dump, what it means, how to recognize them, and how to avoid losing your investment.

Followed that by introducing another very important, and common term, called: ICO-s. I will explain what ICO-s are, and how to recognize the Next Bitcoin, and how to invest in early stage!

Next, I will cover some great strategies and techniques on how to recognize all criteria, that must be considered before investing! This section of the book will include how to identify your investment!

In same cases, it is hard to tell what a certain Cryptocurrency really represent: Currency, Platform, or Application, however, I will explain all in great detail, so you will have an advantage on the market, due to this knowledge.

Then, I will take this knowledge to the next level, by explaining why Portfolio Adjustment most be re-visited, and provide you with examples, so will know how often you should do re-balancing.

Next I will explain how to keep your portfolio profitable at all times, by learn techniques on when you how to re-balance your portfolio, as well if you should outsource this task, by hiring a Cryptocurrency specialist.

Next, I will explain the fundamental differences between centralized & de-centralized Cryptocurrency trading platforms, as well their Pros & Cons that you must be aware.

Finally, I will provide options, what you should do, when your portfolio is dropping. Should you Panic? or look at this situation as an Opportunity? Should you sell? or should you buy more? Don't worry, just keep on reading, and you will learn what moves the market capitalization, and how to recognize the differences between market manipulation & long term success!

There are plenty of books on this subject in the market, thanks again for choosing this one! Every effort was made to ensure the book is riddled with as much useful information as possible. Please enjoy!

# Chapter 1 - Crypto in a nutshell

Cryptocurrency is a digital method of transferring money, amongs many other things. The most popular form, is of course Bitcoin. As of September 2017, there are more than 1100 digital currencies exist.

If you are thinking : Wait a minute Keizer! Are you serious? 1100 Crypotocurency? – that's ridiculous!

Well, you are right! The amount of coins on the market right now, is over the limit, but what can you do. I can't stop engineers to come up with new ideas, and continue developing crypto, neither you. The real problem is that, the large percentage of these cryptos are trying to complete the best ones, or worse, scammers coming up with fake coins to try to make you invest, so they can run away.

My job, is to teach you how to spot fakes, and concentrate only the 10% of the Crypto market, on those coins that are truly represent something, that will indeed, change our life in a very near future.

So, the reality is, that these technologically developed currencies are here to stay, and already part of our life. One of the easiest way to understand quickly what cryptocurrency is, if we take a look at an example. Let's say that you and I want to exchange something between us.

Let's assume, that you are in the US, while I am in England. Traditionally you could send me the goods, once I made a payment to you. Typically I would use an existing bank, or even another known company, known as: trusted third-party systems, such as Paypal, so I could make the transfer to you.

Of course, there are other methods too, such as Apple pay, Skrill, or Payoneer; however, there is always a third party involved, when I would make a transaction to you. All these traditional payment transfer methods, are based on existing financial systems. However,

Cryptocurrency is completely different. Cryptocurrency has its own system, which indeed a software, or to be more specific, it's a protocol that allows me to send a value to you, without involving any other trusted third party.

The software, is running on computers in a peer-to-peer manner, and the system itself is completely decentralized.

Decentralized, by not having a master node, therefore, all nodes have the same control, and same voting power as any other node, that is participating in running the software.

The system is maintained by the computers; therefore it has no central party. What I mean by central party, is that, there is no bank, financial institution, or government running it.

Let's be more specific here, so you understand the differences between a Cryptocurrency payment, versus a traditional banking system payment.

Let's assume that you are in the States, and having a website, selling a digital product, and I am a huge fan; therefore I would like to purchase one of your product, that I can download from your website.

If I made a transfer to you with a traditional method, I would have to use my bank, and the money would go to your bank, however, if we involved PayPal, there would be already three additional third parties, where, my transfer would go through.

I would send you British pounds, which you would receive in US dollars, meaning, additional charges for exchanges; therefore, some money would be lost by the time you would receive the payment.

Banks could take between 3 to 5, sometimes, even ten days, however, if PayPal get's involved, this period could even expand. For example, when I sold my old Rado watch on eBay, the buyer was from Finland, and he has chosen to use PayPal when made a payment.

I listed it that way, by providing additional options for potential buyers, however there was something else, I wasn't aware. From the date of the payment, I wasn't allowed to withdraw my money from PayPal for the next 24 days.

24 days, that's right, not hours, but days. This story is hilarious, because the only reason I sold that watch, is I needed money quick; but I couldn't use it for 24 days, however, by then I was able to get money from another source.

BTW PayPal charged me for 7% because of the nature of the product, a rare expensive watch.

Either way, back to Cryptocurrency, once a transaction has been made, within a maximum of 10 minutes, it will be validated, and you could withdraw the funds right away, using your Cryptocurrency wallet. You may choose to withdraw the funds to your local currency; however, it is not necessary.

There are few differences between Fiat & Crypto, and they are all important. First of all, the transfer is fast, and no need to wait for days, sometimes even weeks.

All Cryptocurrency is running on the Blockchain, where each block gets validated about in every ten minutes. This is the reason I have mentioned the maximum of ten minutes in the first place.

For example, if the last block was validated on the Blockchain eight minutes ago, and I transfer Crypto funds right now, my transaction will be done immediately, and probably gets verified within the subsequent two to three minutes.

Next to point out is the fees. Some Cryptocurrencies are capable of making simultaneous transfers in every second, charging no fees. Some Cryptocurrencies do charge for fees, however, those are usually cents, even in a case of making large transactions such as thousands of dollars.

The reason for low costs when making transactions with Cryptocurrencies, is because, the system is not looking at the amount that's being transferred, instead the size of the transfer.

That is why, if I make a bitcoin transfer, worth of 10 dollars, or 10000 dollars, both times the fees would be somewhat the same.

Next point, is privacy. Some Cryptocurrencies are completely anonymous; however, some others can be chosen to be anonymous, still, it is a huge advantage for many reasons.

Let's assume that I would like to purchase something that I don't want my Bank to be aware. If you or I want to do that, using Cryptocurrency, it can be certainly achievable. For example, you want to help someone, or support a charity that is in Africa, or Asia, using Banks, might be impossible.

Not to mention that using banks, for a purpose of helping others who require, wouldn't get the full amount of the funds, and probably not at the time that you would want it at the first place. There are many more great advantages I could go on about; however, these are one of the most important to note.

In summary, making payments using Cryptocurrencies, are faster, cheaper, more efficient as it's a digital system, no humans involved.

Due to these advantages, many people began to invest in the Blockchain technology, as well in many Cryptocurrencies. The value of Cryptocurrency, is set by supply and demand.

Because, there are an increasing number of people as well companies who are trying to get involved, the value of Cryptocurrencies has overgrown traditional currencies, such as dollars, pounds, and euros. The market of the crypto world is now grown to an entirely different level.

Even governments and large banks are learning about Cryptocurrencies, making sure they are not left out of the new gold, also known as digital gold.

Cyptocurrency was always in the news, explained as the money of the criminals; however, it is now expanded to all levels of people. Regulators are taking especially keen interests, trying to make sure that all Crypto traders have a licence, and Cryptocurrency trading is now legalized in more than 70 countries.

The truth, is that even banks have realized there is an advantage to this structure, as crypto is a spotless efficient system. It is indeed digital money for our digital age. Most people once get involved, will stay involved, and the reason is simple.

Daily payments over the internet, that you and I already doing anyway, using Cryptocurrency, will make our life easier in many ways.

Wall Street, who used to laugh about Cryptocurrencies, are now taking very keen interests on the Cryptomarket too. Alone this year, in 2017 the total market capitalization of all Cryptocurrencies have grown from 18 billion to 170 billion dollars.

This book aims to understand how to invest in the best Cryptocurrencies on the market, and how to recognize a good investment from a bad one. In case you want to learn more about the Blockchain technology, the decentralized ledger, that allows all these Cryptocurrencies to exist, you may read my other books on that topic:

**Blockchain for beginners** – this is a beginners guide to gain same basics on the technology, it is recommended to anyone, even without any experience with Cryptocurrencies.

**Mastering Blockchain** – This is an advanced guide on Blockchain, very detailed on specifically what are the Blockchain attributes, and how they work together.

## Chapter 2 – Tools you start

If you are already invested in some Cryptocurrencies, you may go ahead and skip this part, however, to those, that are completely new to investing, or simply want to know what tools and equipment required in order to buy Cryptocurrencies, it is advised to read this chapter.

First, I would like to point out, that when it comes to crypto investing, it is entirely different than investing in Fiat currencies such as Dollar, Euro, or British Pounds. Fiat currencies are required to have a bank account or even multiple bank accounts for different currencies, however, when it comes to Cryptocurrency, al you need is a wallet.

### Cryptocurrency Wallets

Cryptocurrency wallet comes in many forms, and to purchase one, you don't need to go to the bank. In fact, your wallet doesn't require you to explain your occupation, or race, neither where you live or how much money you make or where you work. Banks would ask such questions when opening an account; however, a Cryptocurrency wallet, is free of charge and anyone can have one or many downloaded in only a few minutes. Wallets are technology, that is part of the Blockchain system, and I am talking about software. Of course, there are hardware Wallets too, however before I would confuse you, let's understand what the Cryptocurrency wallet is.

Traditional wallets are used for holding physical paper money, or coins inside, and once they are lost, you might never see them again. Cryptocurrency wallets in the other hand are software based wallets that have no physical money inside. Crypto wallets have public and private keys. The public keys are the wallet addresses, where you can request to receive Cryptocurrencies.

As the name suggests, public keys, therefore they are visible to anyone. Not only anyone can see the wallet address, but all the founds that the wallet holds too.

Additionally, to the founds, all previously made transactions are visible to everyone too, and they are all registered and can be found on the Blockchain.

The Blockchain is the ledger system that all transactions are recorded, not only the amounts but the sender as well the receivers public wallet addresses. Don't worry the Blockchain does not contain names, only public wallet addresses. Also if you receive for example Bitcoin from someone, the address that you will see from the sender, is the senders public key, aka public wallet address.

I have mentioned that there is another address too, and that is private. Those private addresses, are the private keys, that only you have access to, and no one else. Hence the name private, these keys, or addresses are not recorded on the Blockchain.

Note: If you have read the book: Bitcoin: Invest in Digital Gold, you may skip this chapter, however in case you want to refresh your knowledge on wallets and wallet recommendations, you are more than welcome to read this chapter in full

**Hot Wallets:**

My recommendation is to read this book and learn more before you buy any Bitcoin. However, It's time to understand the wallet technology.

First, I would like to point out that there is a misconception, or I should say misunderstanding what the crypto wallet is. Many people believe when purchasing Bitcoin, it will be stored on the wallet.

This is not true. Bitcoin that you will purchase will be recorded on the Blockchain, in fact, all Bitcoin that ever was mined, and will be mined, are always going to stay on the Blockchain. Bitcoin has no physical presence, and when you are transferring Bitcoin, you don't move the Digital coins anywhere; instead, you assign a new wallet address to the amount of Bitcoin you buy or sell.

Wallets are a software technology, which stores your private and public keys, and interacts with the Blockchain to allow you to access your Bitcoin account. Once you have access to your account, you can send, receive, or directly monitor your balance.

There are many different kinds of wallets, and some portfolios support different currencies, however, some only supports one particular crypto currency.

For example, imagine that you have an HSBC bank card that you can access your Euros; however, you also have another card with Bank of America to obtain your dollars. When it comes to a crypto currency wallet, some wallets can support multiple currencies simultaneously.

## Online wallet

You might choose to have a wallet from Blockchain.info that allows you to have access to your Bitcoin as well Ether.

By the way I have a Blockchain wallet, in fact, that was my first wallet, and I still have it, even it's limited for currencies, I still use it, and recommend to anyone to start with it.

You can have a Blockchain wallet for free on your desktop as well on your smart phone by reaching the website called: blockchain.info

Once you enter the site, you can click on the menu: wallet, then click on: Get started now. That will take you to the registration page, where you can sign up in no time, and buy Bitcoin very quickly.

You can link your current bank account and start to purchase your first Bitcoin. This is how I bought my first Bitcoin, and I would highly recommend it for beginners. It's effortless to use it, and start using your first Bitcoin wallet.

This wallet does not support all the currencies that you might want to invest in the future. However right now the focus is on purchasing Bitcoin, or even before that, to have your first crypto currency wallet that supports Bitcoin.

If you start with a Blockchain wallet, which I highly recommend, do not go crazy and invest all your money. Instead, have a feel for the technology, and start with 100 dollars.

Some people recommend no more than 500 dollars as a first investment, however, if you are a complete beginner, you can do even just 50 dollars worth for the first time. When I have started, I went in with 100 dollars.

When you link your bank account, the options to purchase Bitcoin for 50 dollars, the fees seemed to be higher than buying for 100 dollars worth, so I have chosen to have 100 dollars as my first investment. As I mentioned before, easy does not come with safety, and it is also true with a Blockchain wallet.

For instance, if Blockchain.info gets hacked, the hackers can take control all wallets that reside on their platform.

It had happened before with other trading companies, which resulted in administration; therefore you have to understand, that any online platform exists today is never 100% secured.

Also, once you have a wallet installed on your mobile phone or desktop computer, those devices also can be hacked.

I am not saying that once you buy Bitcoin, you will be hacked in the same day, but there is a potential risk, and you have to live with that. Of course, if you back up your wallet, you have a better chance of being able to retrieve your funds, in case something bad happens.

**Desktop wallets**

Desktop wallets are reasonably secured; however, they can only be used on your desktop computer. Once you have installed it on a particular computer or laptop, that will be your only device where you can access your Bitcoin or other Cryptocurrencies.

As I mentioned before, it is not based in the cloud, and if you regularly turn off your pc, and having an antivirus, you should be just fine. Of course, there are many ways for hackers to get hold of your desktop wallet using various phishing attacks, or viruses from drive by download sites or torrent websites.

Once the hackers would access your private keys, they would be able to access your digital currencies; therefore, it is not recommended for long term use, or even for short term for large sums of coins.

Again, desktop wallets are easy to be downloaded, as well to be used, however still not fully secured.

**Mobile wallets**

There are many different mobile wallets, and Blockchain.info has one as well. However, Blockchain.info provides both, online and desktop wallets. On the other hand, there are some mobile wallets, specifically for cell phones.

Having a mobile wallet is essential for making payments anywhere you go, as your cell phone probably will be with you most of the times. Mobile wallets can provide good security too; however, you must back up your mobile wallet too, same as your desktop wallet.

For example, if you would lose your phone or break it, buying another phone and having backup phrases for your mobile wallet, you can simply back up your new cell phone, like nothing happened.

Again, if you don't back up your mobile wallet, you can lose access to all your Bitcoin forever.

**Cold Wallets**

Cold wallets aka Hardware wallets, those two descriptions are often used both, and really means the same thing. One of these devices, is what you must have if you planning on purchasing large sums of Bitcoin.

These are the best and safest wallets that you can have. Hackers can't do much by attacking it from online, as the cold wallets are keeping the private keys on the hardware, away from the internet.

Some of these wallets, are often look like a USB Stick, and unfortunately, people believe that having a USB stick is the same as having a cold or hardware wallet.

This is not true. USB stick cannot be backed up; neither understands the current market. Of course, hardware wallets do look like USB devices.

Hardware wallets also have to be connected to the internet for sending or receiving funds. However, this would require a special USB type of cable. Any Cryptocurrency that you are afraid to keep it on a mobile or online wallet, you should keep them on a cold wallet. There is no way to someone hack your cold wallet, unless they have physical access to it. However, still they must break your pin code, then must have access to your secret keys that you write down on a piece of paper when you use the device first time.

The only downside, in my opinion, is that if you want to sell some of your Bitcoin quickly, you can not. I mean if you have your cold wallet with you, and there is a desktop pc around, then it can be done in less than 5 minutes.

However, when you have a mobile wallet, transaction can take like 10-15 seconds. Traders always have lot's of Bitcoin on their mobile wallets too in case it's time to sell or buy. I call myself an investor, when it comes to Bitcoin or any other Cryptocurrency.

I do believe that Bitcoin will continuously increase in it's value, and while it does, Fiat currencies will steadily decrease in value; therefore it will provide an additional boost to the value of Bitcoin to grow even stronger.

Hardware wallets are the most secured, there is no doubt about that, however at first, it can be difficult to understand how cold wallets work.
Once you receive a cold wallet, at first you have to create your Pin code. Next, it will ask you if

you are backing up an existing account, or creating a new wallet. So choosing to build a new portfolio, it will generate 24 worlds that you have to write down on a piece of paper which comes with the wallet.

This is also known as a seed recovery sheet. Once that complete, you have to update the firmware. That could take another five to ten minutes.

After that, you have to choose what type of Cryptocurrency wallets you want to download on your wallet. Hardware wallets such as Trezor or Ledger Nano S can support five different crypto wallets at the same time.

Of course you can choose from more than five currencies, however, let me give you an example: Let's say that you want to have Bitcoin, Litecoin, Monero, Dash, Zcash, which is not a problem. However, a week later you want to add another crypto wallet to your hardware wallet, for example, Ethereum, you will not be able to do so until you delete one of the existing crypto wallets.

So, there are pros and cons to it, however to setup a cold wallet you might require to allocate 30-40 minutes for these purposes. Installing a mobile wallet is less than a minute; but you have to respect the security that comes with a hardware wallet. Additionally, if you lose your equipment or get it broken, you might choose to purchase another one, and having those 24 random words you have previously wrote down, you can re-install all your crypto wallets in no time.

A cold wallet is an absolute must to have, especially if you thinking of a long term investment in Crypto, like I do. Even I keep on telling you that you must have a cold wallet, there are few other cons that you should be aware of, or you should prepare to. First is this: Do not keep the recovery sheet at the same place where you keep the hardware. Why is that?

Imagine that you get a burglary, and they find a Cryptocurrency hardware wallet, and next to it another piece of paper that says: Recovery sheet, and below the 24 words that you wrote down. They don't even have to take your hardware wallet.

All they need is a recovery map, as they can recover your wallet, using another device that is compatible with the one you have. As you see, even it is the most secure wallet that exists; still, all can be lost by not taking extra measures.

Let's take another example. This time there is a fire in the building. Again it is good to have a recovery sheet, however, if that gets lost for any reason, well, if you still have the device, you have two choices.

One is to look after that device forever, or send every fund to another device that is backed up, and you know exactly where that recovery sheet is.

You might choose to put it somewhere safe like in the cloud, so you can access it even in the future at any time, however having all those 24 words together not such a good idea, especially keeping them online. So what you can do is this:

Keep six words on Evernote, another six notes on Facebook, another six on Gmail, and another six on Yahoo storage. This is just an idea. However, it's completely up to you but let me tell you another issue with too much security. I will assume that you are men, and you have a wife, who isn't really into the crypto world like you.

Most probably your wife has no clue what Bitcoin is, not to mention hardware wallets, or recovery sheets, and platform exchanges. Let's imagine that you have 10 Bitcoin and each Bitcoin worth around 4000 dollars, but something happens to you. Let's say that you end up in the hospital in the coma. I am sorry that I come up with the worse examples, but my point is this.

There is no sense to save all that crypto money for your son or wife, if they never going to be able to access it. So, because you put the recovery sheet in 4 different safes, and each safe is in various banks, and each bank is in various country, you might choose to have so much security that not only hackers, but even your loved ones won't be able to access those funds ever. What I am trying to explain is that you should teach your loved ones, or you should tell them that you have written down everything step-by-step, how to access those funds, in case of emergency. Probably not everyone will do that. However, I think it's worth to mention to you, so you can decide how to go about it.

**Wallet recommendations**

Some questions that you might want to ask yourself before choosing a wallet:

• Do you need a wallet so you can use it all the time, every day?
• Do you need a wallet so you can keep on buying crypto currency and     holding it for long a time?

- Do you need multiple currencies or only Bitcoin?
- Do you want anytime access, or only sometimes?
- Do you want to get paid in Bitcoin?

Once you have settled and understood your need, you can choose what best fit your requirements, then go for the wallet suits you best. That's being said lets late a look at what I recommend that is a must have.

## Hot wallet recommendations: Blockchain

I have explained already how Blockchain wallet works, and its limitations when it comes to cryptocurrencies. However, this is my number 1 recommendation to those, who are entirely new to Bitcoin and other cryptocurrencies. The platform can be reached on blockchain.info

## JAXX

Jaxx is one of the most popular mobile wallet, and to be honest, everyone's favourite.
Jaxx supports multicurrency, such as Bitcoin, Ethereum, Ethereum Classic, Litecoin, DASH, Zcash, Monero. Jaxx supports 9 different platforms, that including Windows, Apple and Linux desktop, Android, IOS mobile and tablet, Google Chrome and Firefox extensions as well.
Jaxx has an excellent user interface that gives you comfortable easy to understand, easy to use experience.

Having on multiple devices, it synchronizes to another devices too, same as Blockchain wallet does. Jaxx appears to be slow sometimes, however this could be, because it is not open source, and it supports multiple currencies.
Jaxx can be found at:
**jaxx.io**

# Cold wallet recommendations:

## TREZOR

Trezor is a hardware wallet, that currently one of the best in terms of security. It is an excellent cold wallet, especially to store Bitcoin on it. Once you have a Trezor, you will know why it is so secured as it has a screen, that only you can see. I mean, the Trezor screen is not visible on your computer, therefore, hackers can't get to it, and your private keys are completely off line.

The interface is very easy to use, and I would recommend it to anyone. It also have a web interface for easier use, however the screen is built in to provide additional security. If you want to buy it from the source, you can buy one on following the link:
## https://trezor.io/

It's open source, therefore if case you lose it, you can buy another one, or similar device that supports the same functions, then back it up very quickly. It only costs 89 Euros, however, when I bought it, I have also purchased a Cable for an Android phone, and paid for DHL shipping another 26 Euros.

Then I got charged 21% VAT, so instead of 89 euro, somehow I had to pay 145.20 Euros in the end. I am not saying they are scammers, they are not, but you will get charged for postage no matter where you order from, and the VAT itself was 25.2 Euros, which I wasn't happy about either.

Anyways, it is coming from the Chech Republic, and if you live there, you might not require paying the VAT, but I wouldn't count on that. As I mentioned earlier, it is not a scam, and it is a must have, however before you consider buying one, make sure that you are aware of additional charges.

In case you don't believe me, I have taken a screenshot of my order, and as you can see, I have ordered a White color. You have two options when it comes to the color choices: Black or white.

Another issue that was not in my favor is when I bought it, there was a warning message that it would take six weeks for delivery.

However, once you move on, and click on payments, you have options for DHL delivery that is between 3 to 5 days costing 26 euros, or traditional delivery that is 4-8 weeks for 12 Euros.

So I have chosen the DHL, as I wanted to store all my Bitcoin on a secure wallet as soon as possible. Still, after six working days have passed, I have sent them an email asking them what is happening.

They replied that there is a warning message about the delivery that takes at least six weeks. I was outraged, and my point was that if it's six weeks as a minimum to deliver, then why do they still offer 3 to 5 days DHL delivery. In the end I got it after five weeks; however it is a long waiting time, but again it is a must have, so it's all right.

I have survived all the waiting, and my Blockchain wallet didn't get hacked while I did it. Still, I would recommend to buy a cold storage wallet first, only then start buying Bitcoin, especially in large quantities.

## Ledger Nano S

Ledger Nano was no different when it comes to delivery, I had to wait 5 weeks, and few more days too.

The difference was that the Ledger Nano S was only 69.60 Euro and it's already including TAX. Of course, I had to pay for the postage another 16,28 Euros, this time for UPS.

Totalling the full payment of 85,88 Euros. To buy it from the source, just follow the link:

https://www.ledgerwallet.com/r/e101

Ledger Nano had another trick, and that is this: Once I made the payment, I got a confirmation e-mail, stating this:

*Due to a very high level of demand, and in accordance to the delivery terms you accepted at ordering time by checking the confirmation box, your order will not be shipped before September 4, 2017 and then will be delivered to your address 2-5 days later.*

Now bare in mind this was back in June 24th of 2017, meaning I should wait 11 weeks for delivery. I was really sad, and for some reason I can not recall any checkbox when I first ordered one. However, since I have ordered another one for my father, this time I got them checkboxes for sure.

Anyhow, as I mentioned I only had to wait less then 6 weeks for it's arrival, and I am vary happy with it. In fact, I prefer Ledger Nano S, than Trezor.

Ledger Nano is much more beautiful, at least to me, and it's screen is more likeable. It's also cheaper, and the knowledge is pretty much the same as the Trezor. When it comes to the build quality, also the Ledger Nano is what I would recommend. The ledger Nano S has a capability to take a weight of a car, so my first choice is Ledger Nano S

My recommendation is to use a cold wallet, so you can not be hacked, and again, as I

mentioned before, I would recommend to go for Ledger Nano S.

## https://www.ledgerwallet.com/r/e101

Not only because it is cheaper, but it is not that scammy like Trezor, where you have additional VAT fees and super expensive, also the Ledger Nano looks much more better, but in the same time I have them both Trezor as well Ledger Nano S.

**Integration using Ledger Nano S**

Ledger Nano S also compatible with another nine different Cryptocurrency software wallets, that can integrate as below:

- Ledger Wallet Bitcoin
- Ledger Wallet Ethereum
- Ledger Wallet Ripple
- Copay
- Electrum
- Mycelium
- MyEtherWallet
- GreenBits
- BitGo

This is the best wallet I can recommend, however I wanted you to know that are other options too. Once again, you should decide it for your own reasons first, then you will be able to make the best choice for yourself.

Again, there are many wallets that you can use for free right away, and you probably should as your first wallet. However, when it comes to investing, hot wallets are not safe, therefore I would only recommend a low amount like 100 dollars as your first investment.

Anything less then 500 dollars are not that attractive for Hackers, however it doesn't mean that you are safe, therefore only invest in the amount of Bitcoin as much as you are ready to lose, in case you get hacked.

In the meanwhile, you can buy a Ledger Nano S, as it will take few weeks to arrive, and while you are waiting, you can read more books on Blockchain and Bitcoin for better understanding, as well to become more comfortable with the technology. Once you receive your Ledger Nano S, you can start to invest larger amounts, and keep it safe on your Ledger Nano S.

**https://www.ledgerwallet.com/r/e101**

I have both Trezor as well Ledger Nano, and I also know that some people prefer Trezor over the Ledger, however in my opinion, Ledger Nano S is the best. Trezor was released in 2013 and it's design doesn't look so appealing, however Ledger Nano S was released in 2016, with a lot better looking hardware, or at least to me, it is more stylish then the Trezor.

I believe that Ledger Nano S has a way better looking then Trezor, however if you think otherwise, it's fine, but please bare in mind that Trezor is more expensive. In order to buy them again use the following links:

**Trezor:**
**https://trezor.io/**

**Ledger Nano S:**
**https://www.ledgerwallet.com/r/e101**

As I mentioned, I had to wait for both of them, however if you do not want to wait for weeks, you can check on eBay or Amazon where others might sell it too. I have realized that second hand devices are much more expensive. In the same time, it might worth it.

What I mean is first I have tried to buy it on Amazon, even it was more expensive by a third party seller, still I was going to buy it from Amazon, as I trust Amazon delivery, their customer service as well it's fast.

I have Amazon Prime too, and pretty much anything I order from Amazon, I do get it as a next day delivery. So the problem was when I wanted to buy the Ledger Nano S is simply was the availability. Literally Amazon was out of stock too, so, even I was going to pay more for the Ledger, next day delivery would have been awesome, but no stock.

If you are in the rush, or just want to use Hardware wallet right away, like I wanted to, I would advise to check Amazon first, as you may get it lot more faster if they have in stock of course.

## Chapter 3 – Bitcoin ATMs

Some wallets might require you to purchase them online, those cases, you should have an online bank account already, in order to purchase it, however, there are other ways to buy Cryptocurrencies as well.

## CASH ONLY!

But how? Well all you need is to visit a Bitcoin ATM, therefore I would like to introduce you to another great tool, all though not sure where you live, and to find a Bitcoin ATM might be somewhat difficult. Still Bitcoin ATM-s in a rise, and continuously opening new devices all the time.

*Note: I have demonstrated a full chapter on Bitcoin ATM-s in my other book called: Bitcoin – Invest in Digital Gold. However if you want to refresh your memory on the information on Bitcoin ATM-s, go ahead and carry on reading this chapter.*

So, there are machines called Bitcoin ATM-s. They are connected to the internet, as well looking very similar to a traditional ATM-s.

However the purpose of these are to convert Bitcoin to Fiat currency or vice versa. In 2013 the first Bitcoin ATM has been installed in Canada, and it has been a continuous increase of Bitcoin ATM-s ever since. Currently, as of 2017 August, there are 1493 Bitcoin ATM-s around the world, operating in 57 different Countries.

## Types of Bitcoin ATM-s

It worth to mention that are many different types of Bitcoin ATM-s, and some are only operate in one way, however, some has a two-way function.

It also depends where you use the Bitcoin ATM-s , as if you want to convert your existing Bitcoin to cash, you will probably receive the local currency of that country. Other issues that you might encounter is the fees. There are Bitcoin

ATM-s that are operating with no fees, however some others can take as much as 5-10% fees once used. The legacy of Bitcoin is that are no fees, however, there are many operators who paid for producing such machines, as well costs to pay the electricity bills, rental fees, therefore some might charge for a certain fee. It is advisable to check the fees before using one, however the comfort that it provides are extremely helpful.

## How to use Bitcoin ATM-s

In case you wonder how it works, I can tell you from experience that is relatively easy. Let's assume that you want to buy some Bitcoin, using traditional Cash such as dollar. The checklist that you should have is this:

- Smart phone with internet connectivity: Any types of smart phones are ok.
- Hot wallet downloaded on the smart phone: Blockchain wallet or Jaxx
- Dollar bill: Ten, twenty or any dollar bill that you want to convert into Bitcoin.
- Bitcoin ATM near you: You can find a local Bitcoin ATM near you by visiting this link:

### https://coinatmradar.com/

Once you have downloaded one of the wallets I have recommended, or any other Cryptocurrency wallet to your smart phone, visit a local Bitcoin ATM.

In case you think that Bitcoin ATM-s are placed in some dark hidden street, let me tell you that normally the Bitcoin ATM-s are in a public place such as restaurants, pubs, or local shops, places that are many people visit daily.

**Buy Bitcoin**

Once you are there, and happy with the fees that the ATM will operate, do the following:

**Step 1.** Click on the machine's screen: Buy Bitcoin

**Step 2.** Open your Hot wallet on your smart phone, and select: receive. This will bring up your QR code on your smart phone screen.

**Step 3.** Hold your phone to the Bitcoin ATM-s screen, and let it scan your QR Code.
In the meanwhile the ATM will tell you to insert a bill.

**Step 4.** Feed the paper bill to the machine. This time you can feed as much as Bitcoin is available, however if you do it first time, you might try it only with a 10 dollar bill.

**Step 5.** Click send Bitcoin on the ATM. The Bitcoin ATM will perform the transaction, and you might see on the screen something like: Sending Bitcoin. It would take around less then a second.

**Step 6.** Check your smart phone for notifications: You should have on your screen something like: New payment received. Of course it depends on what wallet you are using.

Once you happy with the Bitcoin that you have received, you may also try out how to sell your Bitcoin for cash.

Again, you should check beforehand, making sure that you visit a two-way function Bitcoin ATM, but of course if you only want to buy Bitcoin, than you can visit a one way operating Bitcoin ATM too.

This case let's assume that you are ready to convert your Bitcoin to cash, using a Bitcoin ATM.

Please note that some of the Bitcon ATM-s are the minimum limit is at least 5 dollars, meaning if you only want to sell Bitcoin that worth only one dollar, it might not be possible. It is better to check before, however normally the minimum cash to take out, is at least 5 dollars.

**Sell Bitcoin**

**Step 1.** On the Bitcoin ATM, select: I want Cash. This will bring up the next screen, listing 5, 20, 50 that you can tap on.

Note here: some ATM-s have the same functions like traditional ATM-s where there is an option for: Select other amounts, however some Bitcoin ATM-s have no numbers to create your special amounts, instead, you can keep taping on the listed amounts.

For example if you want 30 dollars, but there is only option for 5, 10 and 20, you can tap on the 10 three times. This will provide you with a 30 dollars.

**Step 2**. Choose the amount you want: For example, select 5, for five dollars worth of Bitcoin.

**Step 3.** Tap on the next screen: Cash out. Once you tap on: Cash out, the ATM will generate a QR code. This will be for you to scan it, using your smart phone.

**Step 4.** On your smart phone, open the hot wallet and select send.

**Step 5.** Hold your smart phone to the ATM and scan the QR code that the ATM has generated previously. Once you have scanned the QR code, using your smart phone, it will generate the transaction that you will have to confirm in the next step. This step, your phone will calculate the amount of Bitcoin that you have to send to the Bitcoin ATM's address.

**Step 6.** On your smart phone, select: SEND. Once you tap on SEND on your smart phone, it will ask you to confirm it again by opening another window.

Here, you will see how much Bitcoin you will send to the address of the ATM.

**Step 7.** Select: Confirm. This will send the transaction to the ATM. This will take another second or two. However once complete, the Bitcoin ATM will come up with a screen: Bitcoin received! > This screen will quickly change to another display where it will say: Dispensing... Please take your cash.

**Step 8.** Take your cash! Check on the Bitcoin ATM below for your cash, and take it.

There are many different Bitcoin ATM-s and some functions different then others, however the process is somewhat the similar.

I have not try them all, however I did try out two of them already and they both were working just fine.

I have tried one that was charging me for 3% and another Bitcoin ATM that had no fee. In the end of the day, for the convenience of selling and buying Bitcoin instantly, and anonymously it's awesome.

In case you are afraid of being hacked online, this is one of the safest way to buy Bitcoin.

Again, I am not sure where you live, however using coin atm radar, you should be able to find one close by to your home or work.

**https://coinatmradar.com/**

# Chapter 4 – Best Crypto Trading platforms

Back in the day you were lucky if you were able to find, at least one Bitcoin trading platform. However, nowadays, there are so many that you cant even count them. The real problem is not to find one, instead making sure that you are not getting scammed on some fake Cryptocurrency trading platform.

I will explain with more details on how to recognize fake websites and scammers on the next chapter, however first I would like to introduce some great platforms that you can use in the future. Please note, this book has been written in August 2017, and I am pointing out the best online trading platforms that are exist today.

Instead of analysing each online trading platform, the question you should ask is this: how to find genuine online trading platforms without getting scammed. In order to find platforms, as a starting point, you should be looking at Cryptocurrencies that are on the market since years such as Bitcoin.

Bitcoin is not a scam, in fact Bitcoin is the strongest Cryptocurrency of all, and of course the first even Cryptocurrency that was created. Bitcoin exists since 2009 January, meaning more then 8 years already on the market and stronger than ever.

Of course you already know that by now, however just to point it out again, if you follow Bitcoin and it's markets, by what platforms it is mainly traded, you should be able to find those online trading platforms that are completely legit. Where to start? The platform that you are looking for is what I already introduced previously that is called coinmarketcap.com

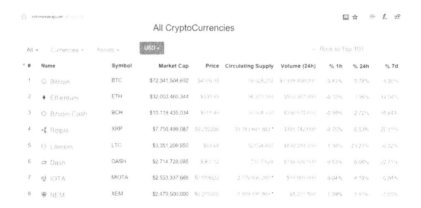

https://coinmarketcap.com/all/views/all/

Once you have navigated to coinmarketcap, you will find all Cryptocurrencies that are on the market. Next if you click on Bitcoin, it will take you to the next page, where you can find more details about the currency:

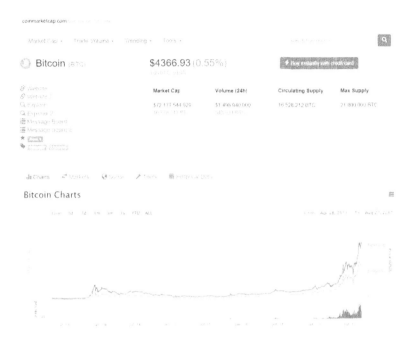

In this page there many information that you can find about Bitcoin, including:

- All the websites where you can find out more information.
- Market capitalizations,
- Additional tools,
- Historical Data, and so on…

It's easy to just get lost here, as there are some great information about Bitcoin, however, our main focus is to find a genuine online trading platform, therefore you should carry on by selecting the menu called: Markets:

By selecting the menu option: Markets, the new window will open where you can find information about:

- All Bitcoin Markets,
- <u>Source – these are the platforms we are looking for.</u>
- Pair – meaning what Bitcoin can be exchanged into.
- Price – This is the price of the Bitcoin on each of those markets.
- Volume – This number represents the percentage of all Bitcoin that is currently on the market; however the number is unique to each platform.

You can also find on the right another button that now shows USD, however, if you click on it, you should see all other currencies that you can trade with when trading Bitcoin.

Bitcoin Markets

| # | Source | Pair | Volume (24h) | Price | Volume (%) | Native |
|---|---|---|---|---|---|---|
| 1 | Bithinka | BTC/USD | $49,390,000 | $4353.10 | 3.30% | USD |
| 2 | Bitirex | LSK/BTC | $49,307,300 | $4498.61 | 3.29% | BTC |
| 3 | Poloniex | LTO/BTC | $49,068,500 | $4378.42 | 3.27% | ETH |
| 4 | HitBTC | BCC/BTC | $43,728,500 | $4357.28 | 2.92% | AUD |
| 5 | Poloniex | XMR/BTC | $39,975,400 | $4323.34 | 2.67% | BRL |
| 6 | Bithumb | BTC/KRW | $35,357,600 | $4378.48 | 2.36% | CAD |
| 7 | OKCoin.cn | BTC/CNY | $33,581,100 | $4358.89 | 2.24% | CHF |
| 8 | Poloniex | LSK/BTC | $31,004,500 | $4436.32 | 2.07% | CLP |
| 9 | BTCC | BTC/CNY | $30,510,800 | $4353.54 | 2.04% | CNY |
| 10 | Poloniex | XRP/BTC | $29,447,100 | $4315.15 | 1.96% | CZK |
| 11 | Huobi | BTC/CNY | $29,156,300 | $4345.86 | 1.95% | DKK |
| 12 | Bitirex | MCO/BTC | $28,997,300 | $4437.16 | 1.93% | EUR |
| 13 | Poloniex | ETH/BTC | $28,678,600 | $4360.14 | 1.91% | GBP |
| 14 | GDAX | BTC/USD | $28,249,300 | $4368.00 | 1.88% | HKD |
| 15 | Bitirex | LTC/BTC | $27,278,800 | $4378.73 | 1.82% | HUF |
| 16 | bitFlyer | BTC/JPY | $22,589,200 | $4371.36 | 1.51% | IDR |
| | | | | | | ILS |
| | | | | | | INR |
| | | | | | | JPY |
| | | | | | | KRW |

At this moment Bitfinex has the most Bitcoin on the market, and it's pairing BTC/USD, meaning you can buy Bitcoin to US dollar or vice versa.

Next by clicking on Bitfinex, it will take you over to the next page, where you can see all the currencies that this site is currently trading with, as well the website on the top right corner, and its Twitter account:

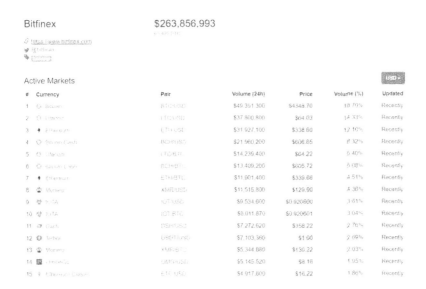

By clicking on their twitter account, you can see how much engagement the site has.

Twitter has all the complaints as well appreciation for the platform, therefore by taking a quick look for educational purposes, is a good idea.

For example, you can see that are more than 56.8K followers, and that can tell you lots too.

If it was a new trading platform, you would probably see lot fewer members, and that case would be advisable not to get involved yet; however more than 50K followers are just fine.

Also you can contact others and ask their opinion about the platform.

Of course this is a genuine platform; therefore you can also visit their website at the link provided on coinmarketcap:

https://www.bitfinex.com/

Once you have reached the platform, you can just register for free, and start trading Bitcoin or other Cryptocurrencies. Bitfinex is one the most respected Online Trading platform for multiple reasons.

Unfortunately, Bitfinex was hacked in 2016 August, and some of the traders also have lost their value, as some of them have left their Cryptocurrencies on the platform instead saving them to a hardware wallet.

The hackers have taken 120K Bitcoin at the time, and even it seemed that Bitfinex would go into administration like some other crypto traders previously, Bitfinex has recovered.

In fact not only recovered, but paid in full to all their traders who have lost their value when the hack has happened.

Even it has taken eight months for them to reimburse all their investors, they could have just closed their doors, however instead they have looked after their investors, and paid back all their losses in full in 2017 April.

Since that event, there have been even more investors than ever, as they have provided an example of trust and long term relationship, when it comes to investing in Cryptocurrencies.

I wanted to show you an example on how to find an online trading platform by only following links, however, as you can see there are many platforms to choose from. Because of that fact, I will cover some other platforms that I have been using previously.

First things first, even Bitfinex have paid back all their investors, which is nice. Still, I hope you understood by now, that no online trading platform is safe. They all can be hacked, and probably will be hacked some day.

Because of the way that Cryptocurrency works, there is very little that police can do, if they might take away all your Bitcoin.

Simply, there is no guarantee that once these platforms are hacked, they will give you back all your losses.

**Coinbase:**

Coinbase is one of the most convenient exchanges out there to invest in, or withdraw money if you live in the US or other supported countries. The website works just perfectly, and you can reach it by following this link:

**https://www.coinbase.com/?locale=en**

The interface is user-friendly, and it is recommended to anyone, even if you have no technical background, you will be able to navigate the site quickly.

Additionally, they have some excellent charts to see when it comes to technical analysis. If you have a problem, you can log a support ticket, and the responses are usually speedy.

Coinbase is also a good place for beginners, however when it comes to authentication, you should make sure that you choose to have a google authenticator, instead of getting verified by MMS. Using MMS to authenticate is one of the ways that your account can be hacked. Therefore it is highly recommended to take the right security measurements.

**Poloniex**

The site can be reached by clicking on the link:

## https://poloniex.com/

Poloniex has countless of crypto currencies; however the highest volume of exchanges are Bitcoin and Ethereum.

This site does not except US dollars; instead you must use Tether, that is another crypto currency which has a purpose of a US dollar. Meaning 1 tether is always around 1 dollar. On Tether, you can find more information on their website:

## https://tether.to/

Tether is a digital value of the dollar. However, some online platforms such as Poloniex do not want to trade with the traditional dollar, instead tether.

Back to Poloniex, they have become so overwhelmed due to too many investors; they are literally super busy for the next few years probably.

What you have to understand is that, when you register on Poloniex, they will check your account, and eventually verify it before they allow you to trade on their platform.

Now that's all ok; however, they have so many new customers recently, that some people have been waiting three months to get verified. However, some others don't even get replied after three months.

The customer service became horrible over the last six months, and many people have just moved on and started trading on other platforms instead.

## Kraken

Kraken can be used for those living in the European Union countries. Reaching Kraken, only use the link:

https://www.kraken.com/

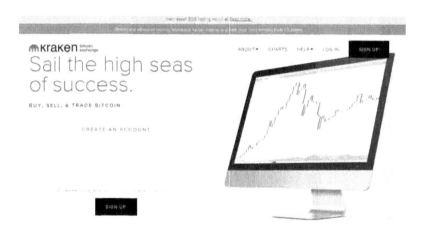

Kraken accepts Euro as well British pounds; therefore it's one of the best choices to Europeans to trade on.

Kraken uses a Tier system, meaning you can start trading as a Tier 1, where you can exchange between all currencies, but account funding is limited to digital currencies only.

Tier 2, is where your daily limit is 2000 dollars worth by both, depositing, or withdrawing, and in both, Fiat currencies as well in

Cryptocurrencies. Tier 3, it allows the same functions as Tier 2, however this time the daily limit is 25000 dollars.

Also worth to mention that if you want to withdraw Cryptocurrency, your daily limit is 50000 dollars.

Same as with other crypto trading platforms, you must get authorized, to get started on Kraken, which has taken me a week; however I have heard that some people have to wait longer than that.

The interface is very friendly, and there are some excellent charts too; however the security is that I want to point out again.

Two-factor authentication is what you want to use. Meaning, you will use your password as well your Google authentication using Google authenticator.

This is what you have to set up, as the system will not do it for you.

You must have a smart phone, and you must download Google authenticator, is a free application, then set it up for better security.

I have covered all those online trading sites I have used previously, and those are all legit genuine websites.

There are many more online trading platforms that are also legit; however, I have not tried them out yet; therefore I can not review them.

All through, you may find other websites through coinmarketcap.com, and they all should be legit.

Still, you should do some more research and make sure that you are comfortable using any Cryptocurrency trading site.

## Chapter 5 – DO NOT GET SCAMMED!

As I mentioned previously, there is a huge industry around Bitcoin and other Cryptocurrencies. Still, many people just beginning of learning about Bitcoin, and most of them who does, want to buy some.

Unfortunately, I have to tell you that hackers are one of the worse enemies when it comes to Bitcoin.

Hackers can hit anytime, at anyone, and they can steal Bitcoin from personal accounts, as well from large online trading companies. Even experienced people do get hacked time to time, and the reality is that most people are only taking security seriously, once they got hacked.

It's nice to own some Bitcoin for weeks, then months, and keep on buying more and more, however, if you do not take extra security measurements, one day you might wake up, for your wallet being emptied out.

Any purchase you have, you must make sure that you use a hardware wallet and save all your values on it. Still, my number one recommendation is Ledger Nano S:

### https://www.ledgerwallet.com/r/e101

I can not emphasize enough, that authorities, such as police or even your bank, will not be able to help you once you get hacked.

There are some trading platforms who might get hacked and will repay you in full, such as Bitfinex did. However, there is no guarantee that they will do it again, in case history will repeat itself.

Now that's being said, as I mentioned there are many new comers to this industry, and beginners who are often having no knowledge, they hear that the value of Bitcoin has been increased dramatically; therefore they quickly want to get some.

Anytime when you or anyone you know is so excited, and want to rush to the market to buy some Bitcoin, please do not do it, and those you know, tell them that is one of the worse things they can do.

Sure thing, you hear that Bitcoin is a good investment, and you have money to invest, so let's do it right? WRONG!

There are so many scammers around the web, that is very difficult to say which ones are legit, especially for those, that are novice to this market.

Once you have the experience, you will know how people with bad intentions are trying to scam people, using Ponzi schemes and other methods.

How do they reach potential victims?

Well, they go where the people are. Those might be Facebook, Twitter, or YouTube, and begin advertise themselves. Trying to make you believe, they will give you a huge profit margin or even make you a millionaire.

They are using techniques such as creating a fake Facebook account, or YouTube channel, as well counterfeit websites.

First of all, anytime you hear things like double your money or making a profit every day, even if Bitcoin market is down, there is something dodgy going on. Unfortunately, there are many people losing thousands of dollars because they fall for some of these scammers or thieves, and there is not much they can do about it.

**Scam alert No1:**

Scammers put pictures of famous people on the website. However, you can not find out exactly who is the CEO, or the link to About Us does not work, or worse there isn't one. There are some occasions where the Contact Us menu option isn't working or once again there isn't one on the site.

**Scam alert No2:**

Unreasonable high return. Anyone who claims that can double or triple your money is 99% scam. In the crypto world, no one can tell you exactly what the market will bring us.

Of course, it is possible to double or even triple your money, however, if someone guarantees that for you, that is possibly a scam.

The value of Bitcoin has increased since January 2017 from 900 dollars to 4500 dollars by the end of August 2017. It is true.

However there is no guarantee for that to happen in the next year, even I really would like that to happen, it might not going to.

**Scam alert No3:**

Unsure the purpose of the company. Sometimes if you can click on the About Us menu option (if it works), and you find that you are not 100% clear what the company does exactly, that could mean a possible scam.

Some might say, that they are trading for you, but then how do they make up for their profit? Well if you can contact the company, you should ask them questions like:

- What is the company does exactly?
- How do they guarantee my investment?
- How do they ensure security?

If you get a response that is a bit wishy-washy, well then probably is a scam. If they don't even reply to you, then again, it is possibly a scam.

**Scam alert No4:**

The particular coin is not listed on coinmarketplace.com. If you can not find a coin that you are interested in coinmarketcap, then you should probably turn around and run away from that coin, as its probably another scam.

If there is a new cryptocurrency, do not go to their own website and start investing. First you should always check coinmarketcap.com, simple as that.

**Scam Alert No5:**

The coin is centralized. I have explained this before on Volume 1 Blockchain book, that once a system is centralized, it is controlled by an individual or a company.

If it has a central Server, that can be monitored; there is no guaranty that they will not go and alter it or to make any changes.

Again, I could go on and on about possible scams, as well how to recognize them, however, if you are a newbie, you might find it useful that, there is a dedicated website for Bitcoin scammers, called:

**http://behindmlm.com/**

This site is a collection of possible scams, where also people reply to one another, explaining experiences, of different websites.

Again, please be extra vigilant, and research on any trading company before you would invest into it. Double check their site and make sure everything works, as well checks out ok.

Also if you can contact them, then by all means do so, and ask questions that might concern you.

Genuine company should get back to you within 24 hours due to time differences, and they would provide you with an answer that you were looking for.

## Chapter 6 – Pump and dump

Now that you know how to avoid scammers, where to buy Cryptocurrency, and what tools you need, it's time to understand how to differentiate Cryptocurrencies, one from another, and identify potentially good investment.

As I mentioned earlier, there are plenty of Cryptocurrencies to choose from, however, when it comes to the choice of which one to invest, it can be a agony.

Don't worry; I am about to reveal the secret, however, let's be fair and say out loud the truth. Most Cryptocurrencies are completely worthless. That's right. There are hundreds of cryptos, and most of them should have never been introduced in the first place.

Some of these fake, meaningless coins have been introduced to the world, for the primary reason in focus to scam people.

The reality is that gambling has no limit, and many who has an addiction to gambling, want to get rich quick. Preferably overnight, however, scammers know that too, and try to leverage on peoples addiction. Because you hear from one source that this new currency called X, will be the future, and make you a millionaire...blabla... Do yourself a favour, and don't fall for it. Instead, look at others opinion too, and wait.

Of course, nobody likes to hold off when it comes to investing, in fact, everyone thinks that investing should be done as early as possible, which is true. However, you should know your reasons, as well your limits.

First of all, do not jump like a madman, and start investing all your money, in fact, you should never invest more than what you are willing to lose. Personally, I invest 5% of my income into Cryptocurrencies, however sometimes, even go as high as 10%.

But, I have my reasons, and that is my monthly limit, period. I know! - Sometimes, I get the idea too: if I had invested twice as much, I would have made twice more!

Again, if it happens to you too, there is a solution, which is quite simple! The answer is this: get over it, and move on.

Make sure you know your reasons, have a limit, and be consistent! Back to identifying potential risks, there is a very famous sentence that you should be aware: *Pump and dump*. Once you get into the world of crypto, and hear someone says for a particular coin that is just a pump and dump, you should investigate it.

In case it happens to be true and the particular coin is really just a pump and dump, additionally and have already invested into it, you should probably sell them all, as soon as you can. However, let me elaborate on this topic, and how pump and dump really works.

**Pump and dump:**

In case you never heard of the term pump and dump, then don't worry, once I will explain it, you will remember it forever. The process is relatively straightforward, however, let's start by saying that is an illegal activity. Pump and dump scams comprise two groups of individuals.

First, there are the performers who falsely increase the price of a certain coin by promoting or endorsing it. They begin to buy the fake cheap currencies, and once ready to dump them, and they build up a hype.

Once the hype builds up around the fake coin, the trading volume began to increase, and its value goes up. The person or group is also known as both, the pump and dumper. Once the coin hits the wanted price, the performers sell all their coins, and people begin to panic, meaning they all too, start to sell, and dumping their coins to the market and making the price to drop.

## Advertising the coin:

To notice a coin that's being in the picture for a pump and dump, it's relatively easy. You'll often see buying patterns, by the price is falling and rising just considerably each time the performers buy, stacking up on the cheap coins without showing too much attention.

Once they've to purchase the coins, the performers going to the forums and chat boxes, and talk about why this new coin is so good of an investment. They'll use several different accounts; however, there might be more people involved in this illegal activity, to make it more believable.

They will continuously talk about the coin, until there is hype and people begin buying it. Once people fall for it, and start buying it, the pumping happens.

These scammers might even go as far as tell you what the best platform it to buy, and making all the followers go to that platform, and purchase as much as possible. This situation pushes the hype even further, and more people begin to buy.

Once the coin hits the wanted high point, the performers start selling their coins, however, not all the coins in the same time, but this is the time when you should see that the dumping is about to begin.

You might see this happening only in a few seconds, however good pump and dumpers, would take their time, and do it for hours, if not for days.

These performers would begin to sell only small amounts of coins, as fast as they can, making sure the price is not going down quickly, doing so until they sell all their coins.

Once the performers are off the market, panic begins, and people start selling their coins too, and this is the peak time for the dumping process.

As the price, as well the volume is going down, people begin to realize that their coin is stuck on the market, and no one buys them anymore.

This is when they Panic big time, and start to sell below the market value, just to get rid of all the a coins. Once this happens, the currency becomes completely worthless.

The scammers, of course, are getting out of this, at least doubling, if not tripling their money, and those who were following them, only losing their money.

Successful scammers might even announce publically that they have made a mistake, and lost a lot's of money, because someone have scammed them, while, they are the real scammers.

Nobody will know for sure who was behind all that of course, however, sometimes these people are just too visible to see that something is going on.

If you are capable of spotting a crypto coin that's getting ready for a pump and dump, you may purchase cheap coins yourself.

If you can grab up some coins before the pump starts, you can make money if you're not greedy. It is not hard to make decent profit in few minutes if you've spotted the signs early enough.

If you arrive late to the party, and the coins has already begun being pumped, but still in the early stages, you still can make a profit.

Of course, the risk will be bigger and your profit will be smaller, however, if you enter and walk away fast, you should be able to expect a modest profit.

You might find people, offering to participate in such pump and dump games, even offering Bitcoin in exchange; however, this is illegal, and I highly advise you to stay away from such games, especially if you are a beginner.

Not to mention that some of these scammers go as far as making you believe that they will pump and dump with you, then it might turn out that the only person doing the pump and dump is you.

Stay away from games like these. Cryptocurrency has it's real power in the technology that it uses, not scam artists; however, I wanted you to have a bit of understanding when it comes to entering the crypto market.

Pump and dump are a very risky play, even for those are expert in manipulating the market, therefore do yourself a favour, and do not participate.

Instead, spread the word about pump and dump coins, and help people not to get scammed.

Either way you decide to go about it, there will be hungry wolfs in every market, and they always try to leverage on the week, and those that are inexperienced on these playgrounds.

**Chapter 7 – ICO-s**

Before jumping into how to identify your investment type, I thought that you might be interested in the term called ICO.

Everyone talks about the next ICO will be this, and that and the other, but what an earth ICO really means! In case you are aware of ICO-s, you might skip this chapter, however, if you are a newbie to the crypto market, you should be aware what an ICO means, or to be more precise what are the ICO-s.

ICO stands for Initial Coin offering. ICO sounds like an IPO. However, IPO-s are Initial Public Offerings. IPO-s are regulated, while the ICO-s are not regulated, or at least not yet controlled. However, ICO-s are happening when a particular company raises funds, to distribute a token to their investors. It is somewhat similar to crowd funding, which is also an excellent way to engage with the community, and allowing people publically, to invest into the companies funds, in exchange for distributed shares.

You have to think of it like this: You have an excellent idea, something that you can describe and believe you can achieve with dedication and hard work.

Something that you would be proud of. Something that you are truly believe that can help out people in the future.

Something that possibly has a potential to change the lives of individuals, in a better way, by thinking or doing something different.

Different, that would be involved, terms like: faster, safer, cheaper, or easier. It could be a solution for an existing problem, something that is not obtainable for everyone.

However, you might have an idea that can solve that issue, by helping those who needed such service.

Either way, to proceed, you will need more funds. Therefore, you would present your idea publically, hoping to find investors who can help you to succeed and achieve your goal.

Again, this is an excellent opportunity to continue your dream, which eventually will help others by providing a solution for an existing problem. In a perfect world, that's all there is what an ICO is.

However, we do not live in a perfect world. Therefore, you should not believe anyone, especially to those don't seem to be so excited of these potential dreams to come through.

## Downside of ICO-s

There are so many ICO-s, it's just crazy! Every day there is a new ICO, and the reality is that most of them just fraud. But how they work?

Well, the company should publish a whitepaper, that explains how this new Cryptocurrency would work, as well what problem it solves, how it solves it, why it would be beneficial, and to who, and so on.

Normally, these whitepapers should be a technical analysis of the product, preferably written by the original developers.

If you are familiar in Information Technology, perhaps, you are a programmer or software developer, then you could spot right away, if the whitepaper was written by IT professionals, or just some scammers.

Anyhow, in case you do not have an IT background, spotting the scam might be difficult, still what you should know, that ICO-s should not be too easy to understand.

ICO whitepapers, should not be like marketing materials, instead, technical documents.

If you do come across with 10-20 pages of sales page, or marketing booklets, I will warn you now: perhaps, you should move on and find another ICO.

A whitepaper should represent the underlying protocol of the new coin or token, nothing more.

**ICO example**

There are hundreds of ICO-s where only scammers are trying all sort of strategies, such as pump and dump, pushing up the value, sell them off for a high price, then the so-called token developers could disappear forever.

This is one way of doing business, however, personally not a fun of scammers, performers, who are just lying, and leveraging on anyone they can.

One of the first, and most successful ICO that ever happened, was an example of Ethereum. Vitalic Buterin, a Russian programmer, was 21 years old when he invented a platform, called Ethereum.

He has created his whitepaper, published it, then promoted it, so he can gain investors into his idea. The offer was based on the facts of: those who want to participate, should exchange bitcoins for ether.

Those who have been involved, have traded their bitcoins to ether; therefore lots of ether was bought, and that supported the platform to grow, and it does ever since.

**Issues with ICO-s**

First of all, ICO-s could be offering tokens, or Cryptocurrencies. Once you have understood what is the offer, and ready to invest, you should realize something important, which is the storage. As there is a new ICO that offers shares, you might be able to buy x amount of that coin or token. This is great, and if you are capable of investing into something that can grow dramatically, you can see 10x, 100x, even 1000x return in the future.

However, as this new coin just about to make it to the market, assuming there are some already in existence, you might find it hard to find a wallet that supports it.

Because the coin is still new, it is highly unlikely that any crypto trading platform will support it on their wallet. What these new companies normally do, is they come up with a temporary solution to keep your coins on a particular website.

The problem is this: these websites are usually under the control of those, who proposed the ICO in the first place, therefore investing in heavily is a huge risk.

They might be having a good idea, and honestly believe in their project, however, they could run out of money, change management, or simply make mistakes by having vulnerabilities, that hackers can exploit.

It has happened before that hackers began to copy the initial idea, and created additional coins, even before the real company would have done so, making the currency worthless.

Times like these, early investors, who supported the original idea can lose all their investments, as well the company could face the potentials of delays, or worse: stop the project.

Again, ICO-s might be just a pump and dump, however in the a case of Ethereum, Golem, ICONOMI, was certainly not the case of scam.

**Right ICO Investments**

Most people believe, the technology is what they invest into. That's not very good advice, of course, it's only my opinion.

You can invest in anything you want, but get this: Technology exist and evolve, for one reason, and that reason is us. That's right; it is up to us: humans, if we support the technology and carry on developing it or leave it behind.

Technology does not produce itself, or at least not yet, therefore, your best chance when it comes to investing, is to invest in people.

It is the people, who come up with great ideas, and good investors know it. Successful investors, look at the idea what people have, and support that, in exchange for a return.

You might become an inventor, however, if you have a good understanding what the good idea can be, something that can indeed come through, you may go ahead and invest in that idea.

Furthermore, because, there is an excellent idea that XYZ came up with, still, you should investigate it further by understanding what qualifies the person for being an expert in his or her field.

This is where you should look up the history of the individual, and read his or her previous projects, including past success as well failures.

Additionally, you should identify if the inventor is capable of doing what he or she does best alone, or by managing a great team.

Next, in case of a team, you should understand if the team is managed properly, meaning they are creating an excellent teamwork, as well capable of training others, or they are just stubborn people, who want to do everything by themselves without listening to anyone's advise.

These are all essential elements when it comes to a visionary team, whose aim is to become successful.

You don't want to invest into a team where everyone has their idea, and they can not work together; therefore it's something that you should pay attention to.

Next, look at the community that follows them. Are they open-minded, and positive thinking people, or just a bunch of investors who want to become rich overnight. This would be measured by other researches, such as looking at their social media websites, like their Twitter, Facebook, Blogs and so on.

All these sites should be open for public view, and you should find them, and see how they are engaging to their followers.

What is the problem they are trying to solve, how they solve it, and can they explain it in simple terms to those, who would require such solution? Now that's a bit of a tricky one; however, I hope you get the point.

Let's imagine that I am 100% confident that I can solve a certain problem for you. You might not be aware of it, but today is the day when I am telling you that I am indeed capable of solving your problem.

So, you might be interested, and you question me, how exactly can I do that, because you don't understand. If, I begin to talk about cryptography and Diffie – Hellman key exchange that is un-hackable, and that's the future, blabla…

Would you consider me as a problem solver, or just a nerd who probably knows what he is talking about?

The Blockchain is very difficult to understand, as well difficult to explain. The concepts are covering multiple technologies, that are working together.

However, Blockchain in simple terms is solving the problem of trust, by eliminating trusted third parties, operating in a decentralized manner, on a peer-to-peer network.

Of course, I could write multiple books on the Blockchain technology itself, which I already did; however, there are always ways to explain even complex technical inventions in plain language.

In the case of ICO-s, where the intentions are not clear, neither the implementations, there is a possibility that you should stay away from those.

## Ahead of its time

This is awful to mention but happens all the time. There are some excellent ideas out there; however, some are just ahead it's time.

Now that's fine, however, those, who want to support the idea, but the timing is wrong, can quickly lose their investment.

If the market is not ready for the motion, it will not sell, period. Yes, we all heard of Einstein, who has predicted certain things and believed that it will happen in the future.

Still, nobody paid much attention to it, because it was to early, and people were not ready for the idea.

Here is another example. If you were to come up an idea of an international video call, using an application called Skype.

Skype, which will be freely available to download from Play Store, on an operating system called Android on a Samsung cell phone, using 4G wifi signals, in the 1980-s, most people would believe that you are crazy, and probably no one would fall in love to such idea.

In fact, it would have been likely to sound like a scam to invest, not developing it.

Nowadays it's part of our daily life, and we don't even think about it, just using those technologies combined. However, back in the 80's, the market wasn't ready for such invention; therefore it would have been failed.

To find a good ICO, you must consider all those facts I have mentioned, as they each have their part, and represent the products, as well its market.

Because you, and the inventor believes in it, it doesn't mean that the market believes in it too. This is something that you must keep in mind.

## Chapter 8 – Identify your investment

You might be thinking: OK Keizer! Enough of this teasing, and show me how to make money! Right?

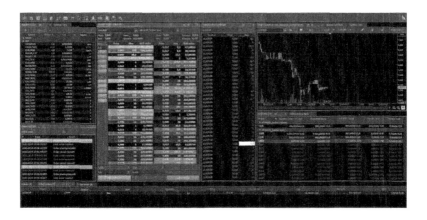

Fine, I will now reveal the secret of spotting the next Bitcoin, or at least how to spot the real potentials when it comes to Cryptocurrency. Instead of listening to others what they believe in, and hoping it isn't just a pump and dump coin, such were the Onecoin, you will quickly realize that all you need to succeed is you.

That's right; you don't need anyone to tell you if a coin has good potentials or not, all you need to do, is research. I am serious now; you should not listen to me either! What I mean is this: If I say that Bitcoin is the future, well it certainly is; still, your job is not to believe me, but do your research, and understand it until you are certain about it. You should never point your finger at anyone, either it's a right or wrong investment you made. Once you do invest, it must be your own decision alone. You must keep your head up, and be fully aware what is it that you are investing to.

Now, let's move on and see how to find that so called next Bitcoin. As I mentioned it earlier, there are hundreds of Cryptocurrencies out there, and the large percentage of them are just rubbish. It would be a waste of time even to look at them.

However, real investors might take a look at 20, sometimes, even 50 Cryptocurrencies before they would choose one. I am about to reduce that process, so you can save time for yourself, and allow yourself to focus on what most important. This chapter is called: Identify your investment type, so, let's dive in.

## Condition

Let's assume that you have found 20 good Cryptocurrencies, each has good potentials, and you want to start investing. However, you are not sure, how much percentage of your money you should invest into each one of them. When it comes to investing, there are three condition types that you should be aware, and look at.

- Idea
- Idea + Implemented (An idea that is already working)
- Implemented with customers (Working, as well people using it)

Let's look at each of these, and understand what I mean by that. An idea is typically an ICO. An idea, however, it's a non-existing project; therefore there is no Cryptocurrency yet.

Moving on, some good ideas already have been implemented, and already working; however, it has no customers yet. 90% of these are on the market, however, let's move on once again.

There are some great ideas, that not only been implemented, and working, but also have customers too.

These are the currencies that take 10% of all the whole market. For example, Ethereum, and Bitshares, are both a good idea, already working, and both have many customers.

However, the market is not interested in neither of those; therefore the prices have been dropped dramatically. What manipulates the value of

Cryptocurrencies, is something that you should always remember:

**Supply and demand.**

I am specializing in Cryptocurrency investing, however any business you look at, there are always some basic ideas that are easy enough to come up with, and that's an idea.

It is easy to come up with an idea, however, to implement it, takes time, money, courage, continuous determination, and hard work.

You have to understand, to produce a product or software, and to make it work in a user-friendly environment, is a huge jump from *just having an idea*.

Even though the product has been created in an excellent working condition, to inquire customers, and begin to sell it, it is another huge jump from just having an idea that is working.

What you have to understand is your investment. If someone asks you, or you may ask yourself, what is it that you have invested into, you must know it.

Are you investing into an idea? Are you investing into an idea that is working, but has no customers yet?

Or lastly, are you investing into an idea that is not only working but already having customers.

This is very important, in fact you might find it that is the best advise in this book, however, let's move on.

**Types of currencies**

There are three different types of Cryptocurrencies:

- Money of Exchange
- Platform
- Application

There are Cryptocurrencies that its core usage is to make payments. They might be used for other than that, however, mainly used for something in exchange.

Another Cryptocurrency type that's a bit different, normally its main advantage is, to provide a platform for many other applications or payment systems.

Lastly, there is another kind of Cryptocurrency, which is used for a precise application.

Again, just so you remember, there are currencies, which might be utilized for all three above mentioned; however, you should look at its core usage when identifying the product.

For example, Ethereum can be used to make a payment, in exchange for a product or service, or even to exchange to another Cryptocurrency, however, it's main function is the platform; therefore Ethereum should be considered as a platform.

|  | Money of Exchange | Platform | Application |
|---|---|---|---|
| Idea | ATBcoin | EOS | Filecoin |
| Implemented | Decred | Stratis | Siacoin |
| Implemented with customers | Bitcoin | Ethereum | TenX |

If you look at the first lines, where the ideas are, you can see that these are the ICO-s. ATBcoin is one of them, followed by EOS, and finally Filecoin.

Looking at the second line, those Cryptocurrencies have been implemented, and already working; however, they have no customers yet, therefore no one using them.

These types of coins already worth more in their value than the first line of coins.

Bear in mind that I have created this matrix, in early September 2017, therefore if you are reading this book at a later date, you might find that some of these coins will have customers, however some might lose their costumers.

Either way, the fundamentals will not change even at a later date, so let's move on. Looking at the third line of coins, these cryptos already have costumes, and these are already in use.

Bitcoin is the most stable, and mostly used as a payment method; however the Ethereum is mainly used as a platform, and finally, TenX is one of the most known applications that already have customers.

That's being said; there are some occasions when it is challenging to understand where to categorize an individual coin. Take it for an example of a cryptocurrency called: ripple.

Ripple, which already has costumes, still it's hard to decide if it's a platform or application; however, the ripple is certainly not money for exchange!

You may find it difficult to know where to categorize an individual coin, however, when you begin to invest in crypto, it is imperative to classify those coins that you want to invest into.

In case you want to know why it's important, then let me tell you why, by looking at the next slide.

|  | Money of Exchange | Platform | Application |
|---|---|---|---|
| Idea - 5% | ATBcoin | EOS | Filecoin |
| Implemented - 20% | Decred | Stratis | Siacoin |
| Implemented with customers - 75% | Bitcoin | Ethereum | TenX |

As you see I have added 75% to those coins, I call: Implemented with customers.

Now, you do whatever you wish, but I Invest 75% of my money to those coins that not only a good idea, implemented, and working, but also have customers, preferably since years already.

These types of currencies mostly can be found within the top 20 - 30 Cryptocurrencies when you look at the list on the website called

www.coinmarket.com

Next, I invest 20% of my portfolio into Cryptocurrencies that are already working, but having no customers. Finally, 5% of my portfolio is invested into those coins; I call a perfect idea. These are mostly ICO-s, and there are only very few out there.

However, I just mentioned that I only invest 5% within ICO-s, and only those that are an excellent idea, however I would like to add to this.

Yes, there are some great ideas, however, if there is a new ICO that will be some money of exchange, and not a platform or Application, I will pass.

Meaning, I do not invest any money into any new payment system, and my reason is this.

There are so many payment solutions exist already, that even if it's a great idea, it would typically have a hard time to compete with existing payment solutions, such as Bitcoin, Bitcoin Cash, or Monero.

Ideas for payment solutions are over exaggerated, and there are so many top Cryptocurrencies for that purpose, that coming up with a new idea on that market, would probably fail.

Of course, this is only my opinion. Therefore, 5% of my portfolio is invested into new ICO-s that are platforms or applications only.

# Chapter 9 – Platform & Applications

I would like to show you some great platforms; however you might get surprised, that even there are some excellent ideas that already in working solutions, as well having customers, still have no real volume.

## Bitshares

The product of bitshares is itself, a complete decentralized Cryptocurrency exchange. You can find their platform, using the link below:

https://bitshares.openledger.info/market/OBITS_BTS

They plan, to complete against the Wall street, and take over within the next few years.

They already achieved to make more than 3000 transactions within an every second; therefore they have the same capabilities like VISA Cards.

This is an excellent idea, and they have some very cleaver team working, making sure they have no issues by creating a decentralized Cryptocurrency exchange.

Unfortunately, there is no volume, therefore only very few investor participated within the idea.

However, in the future, this could change, but for now, it might be not the best investment, and the reason is that, there is no demand for the product right now.

## Ripple

This is a platform that is used by the banks for currency exchange. It is centralized by those banks, that are part of ripple.

They do have many customers, and the product is already working. However, the value of ripple is shallow.

As you see, they have one of the biggest banks behind the product; still, there is no demand for ripple, its values are not moving in any direction, accept sideways.

To visit ripple and find out more, follow the link below:

## https://ripple.com/

## Storj

Storj has no customers; however, there are tens of thousands of people provided their computers for the Storj network. They aim is to create a decentralized network for the purpose of storing files in the cloud.

The following map will show you all those computers, and nodes who participated within the Storj network.

Even they have no customers yet, still, there are huge potentials with the idea, and the bad news is that they only have received 40 million dollars yet. Therefore, it will be complicated for them to carry on working on the project.

In case you want to visit their website, follow the link below:

**https://storj.io/**

## Filecoin

Filecoin, only an ICO yet, and their aims are somewhat similar to Storj; however, there is no product yet.

Again, this is an application that is only an idea, however, when the ICO happened, they have received 200 million dollars within 60 minutes.

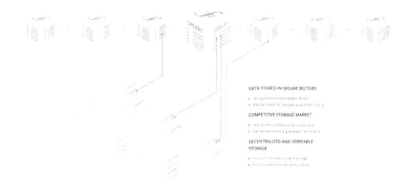

This is just one of those situations that if you were late for the ICO, you would have no chance to participate.

Again, there is no product yet, only an idea; still, Filecoin has set up a world record of most successful ICO-s ever happened, by collecting 200 million dollars under an hour.

To be fair, behind the Filecoin project there is a man called: Juan Benet, who has been known for his previous projects, such as IPFS.

Those who have done their research, know that it is a <u>must to invest</u>, however when you look at my matrix, I would only spend within 2.5 to 5% to ICO-s that have no products yet.

To be fair, I couldn't invest into Filecoin yet, but I will probably invest 10% once it has a working product on the market.

To visit Filecoin website for further information, follow this the link below:

**https://filecoin.io/**

## Chapter 10 – Adjust you portfolio

Hopefully you have understood how to use the Cryptocurrency portfolio matrix, and you will take actions too; however, the market often changes, therefore you should be aware of potential losses, or gains.

Of course, if some your investments are increasing in profit, you don't have to worry; however, there are times when it happens otherwise. Let me elaborate on this.

As I mentioned before, some of the ICO-s might be only good ideas, however, they have no working products yet. As time passes, those great ideas can become a working solution and the actual product could potentially worth lots more than before.

Somewhat the same can happen with a product, where I invest 20% of my portfolio, and suddenly creating a large community; therefore it can began to have a huge customer base.

This is another great news when it happens, as you can experience growth of values within those Cryptocurrencies if you have invested.

In the other hand, you may choose to invest 75% of your portfolio in a product that has a massive customer basis, however suddenly begin to lose its customers or at least a significant percentage of them.

That's being said, all three different types of portfolio investments can decrease in value.

Either way, first of all, do not panic. Instead, learn now, that your portfolio should be adjusted in each of every month.

I say every month, however, there are times, when the market moves quickly, therefore you should take a look at each of your coins that you have invested, and analyze them all, one by one.

This will help you to know how you must re-adjust your portfolio.

This is important, as certain conditions can improve or ruin some crypto coins, therefore it's vital to rebalance it, or at least double check your portfolio at minimum once a month.

You might choose to outsource this task, and hire an accountant to manage your portfolio.

There are individuals, as well companies who already begin to consult and help people or businesses to manage their Cryptocurrency portfolio.

One that comes to mind is called: Colombus Capital. You can reach their website by visiting the link below:

**https://www.columbuscapital.com/**

They are investment specialists exclusively in Blockchain based assets. I have never used them; however, I am aware, they can help to choose profitable Cryptocurrency basket, where, you can decide to invest in multiple coins in the same time.

They do charge for 3% management fees, however in exchange, they do sit down in every month and re-balance your portfolio, making sure that you make the most profit with your investment.

It is optional, however, if you are one of those who wants to invest in multiple Cryptocurrencies, and want to make it profitable in the long term, but having no experience on the market, or simply has no time, it might be something worth look into.

In my opinion, hiring a professional is always more beneficial, not just because you can have more time to enjoy life, but there is something else that I will explain now.

I have mentioned in the earlier chapters how you should adjust your portfolio when it comes to investment, which was as follows:

75% to those that are working and having customers already, 20% of those products that are working already, but have no customers, and 5% is those ideas, that probably still an ICO.

### This is not all!

You should know, when it comes to investing in crypto, or on fact into anything, there are always more things to consider.

For example, let's imagine there are two different products, that each has customers already. However one of them has only one customer, but the other has hundreds of customers.

Furthermore, the product that has only one customer, that customer might be Microsoft, which is operating all over the world, while the other product has hundreds of customers, but those are only individuals or small businesses.

Let's move on to a product that 's already working but has no customers yet. Imagine there are two different products, that each is already implemented, but have no customers yet.

However, one of them is planning to have few customers within the next 3 to 6 months, while the other product is expecting to have thousands of customers within a week.

Moving on to the idea, where I have mentioned 5% of the investment that is reasonable. Let's imagine there are two different ICO-s, and each has no products yet.

One of them sounds very promising to you, but you never heard of the inventor yet, while the other doesn't sound like a good idea to you, but you also learned that the creator of that project is Vitalik Buterin.

As you see, because it's an ICO only, it doesn't mean that you should only invest 5% of all your investment, in fact, you might consider to invest 80% to a certain ICO, and your reason can be just because you know who is that you are investing into.

My point is that, besides the 75+20+5% investment matrix, within each section there are so many choices to be made, and you just can not know everything.

Well, you might be able to research all the time; however, it takes time, in fact, lot's of times to follow the development plans, as well, understand the market around those people or businesses that are in question.

So, when it comes to Cryptocurrency investors, those who are specialists, believe me, they have to work hard all the time, to keep up with the market daily.

Portfolio managers are checking the values of all currencies all the time, and each month they are making the required adjustments, to be most profitable.

However, if you do it for yourself, you may not require doing that in every month, especially if you only have invested one or two Cryptocurrencies.

In case you want to invest in a larger crypto basket that contains 10 to 20 coins, than you really should look into rebalancing it, at least once a month.

Either way, you planning for rebalancing once a month, or once in every second month, when you do so, you must make certain decisions.

You must look at two thinks when rebalancing your crypto basket.

Firstly, the change in the value of each coin, and secondly, you must research on what news have been published on the market on those coins, that you have invested.

Those two factors must be considered when rebalancing your portfolio.

Once you begin analyzing that some of your currencies aren't made any profit, in fact made you lose money, and the news on the market is that those coins are completely useless, you must dump them.

## Ripple VS Bitcoin between May to September 2017

As I mentioned in the earlier chapter, there are some great ideas, and some of those already working and having customers too, still not making any profit.

Those must be dumped, simple is that!

When you are making a business decision, there is a place for emotions, such as you like the product, or you believe that will change the future, and that's fine.

However, if the currency stays in the same value for weeks or months, while Bitcoin and Ethereum are continuously growing, you must dump those coins, and forget about them.

For example in May 2017, Bitcoin was worth of $2400 dollars, while Ripple was worth of $0.40.

Since that three months later Bitcoin hit all time high of $5000 dollars, and Ripple is still $0.30 dollars. I don't have Ripple, not because it would make any profit, but because it's centralized.

To my mind that shouldn't even be called Cryptocurrency, just Fiat currency, but that's just my opinion anyways.

If I do want centralized currency, then I can just go and get some more dollars, pounds, or Euros, however I am fine for now.

Anyways, back to maximizing your profit! There is a tendency, when the value of Bitcoin goes up, every other coins are growing too; however some do not (ripple), and those they don't, probably should get rid of from your portfolio.

Sometimes it's only visible after few months, however, when the time comes, you must make a decision, and take action on it.

Again, it doesn't matter how much you like certain coins if the market doesn't like it. If there is no volume for it, and no one wants to use it, you have to go with the market and dump those coins.

Once you got rid of those coins had unacceptable profit margin, you must replace them with those have made your profit.

When you are investing, you must always maximise your profit, and minimize your risk, simple as that.

**Summary**

As you see, there are many things that you must consider when it comes to portfolio adjustments. To be a successful investor on the crypto market, you must do a thorough research, possible all the time, and make tough decisions.

It is, even more, difficult, when you want to maximise your profit, by investing in 10 to 20 different coins. This is the reason why found managers will always have jobs.

There are very few people, who have time to do thorough research on every Cryptocurrency, and most people don't want to have 20 different types of wallets and look out for each currency all the time.

Instead, what most people want, is to keep all their currencies at one place, also known as one-stop-shopping, where, there is a centralized place for the full portfolio.

For this centralized place, you might hire a fund manager, who can manage your entire portfolio.

I know that some people don't want to hire and pay someone to do that, instead watching the news do continuous research and do all the adjustments themselves. Here is what I can tell you.

You have to see for your self, and probably give it a go, using both ways! This way, you can compare who does it better; yourself or someone you hire for these task for an individual fee.

For example, you might choose to start investing with 500 dollars, and I would recommend to start with $400, or $600 instead, and do the following.

For example, if you choose to start investing with 600 dollars, hire someone to do a crypto basket for you with the most profitability, and invest into that model 300 dollars.

Then use the other 300 dollars, and do it yourself.

Then within three months, see what model made you better profitability, and then invest in that model as much more as you not afraid to lose.

Of course, you might consider the time too, that you have to spend for all your research, and divide it by the hour that you would pay for your own services hourly, however in the end, it is entirely up to you.

## Chapter 11 – Trusted third party VS DIY

As I mentioned earlier, you can choose to hire a Fund manager, or a company who can manage your portfolio. As an important side note, there are a couple of things that you must be aware.

First of all, Bitcoin has been build on a decentralized peer-to-peer network, and this is what makes it so compelling, however, once you choose to keep your full portfolio on a trusted third party, such as a fund manager's account, or even any other crypto trading platform, you are taking an unnecessary risk.

Once you choose to keep your portfolio on a trusted third parties platform, you will have a free wallet, that will provide access to all your Cryptocurrencies; however, your private key will not be yours. Instead the exchange or the fund manager is who will control your private keys, and not you. Please, do not forget that. Second, when you are using a trusted third party, they will require you to provide your passport, ID, address, phone number in order to register with them, and this is not because they want your details, but because they must do it by law.

The law is for anti-money laundering purposes mainly; however, there is a possibility that IRS will be interested too, when the time comes, and of course this has happened before.

Thirdly, it applies to both, fund managers, as well exchanges, that they are wide open for any hacking attacks, at all times.

The hackers point of view, any Cryptocurrency trading platform, is a paradise. To hack your private key, and steal your money, is not only difficult, but a time waster for hackers.

Instead, they target fund managers, and large currency exchanges, those that are centralized, meaning easy to hack.

As you see, these currency exchanges have thousands of peoples portfolio. Therefore our digital age is the best time for hackers to utilise. They don't need to steal information, or identifications, then take the risk of selling those on the black market, instead they can steal bitcoins. And of course where the most bitcoins are is? You guessed it – they are all there, at the exchange platforms.

And the worse thing is that, all most of those trading platforms are unfortunately having centralized databases.

I just can't recommend enough to have your own wallet, so you can become your own bank! Simple as that, do not take unnecessarily risk, there is really no point. Instead invest first in a hardware wallet, and protect yourself, protect your funds.

Learn how to do it, believe me, once you get a hang of it, you will never forget it. Still my best recommendation is to go for a Ledger Nano S!

The Ledger, is simply to use, looks great, the strongest you can but, and comes in a cheapest price! In order to purchase the Ledger Nano S directly from the developers, go ahead and use this link:

https://www.ledgerwallet.com/r/e101

Finally, I would like to mention that these companies are running their businesses because of their computers and servers are allowing it, and there are times, when they have to update their systems and stop their services for maintenance purposes.

However, if you choose to have your funds in a decentralized Cryptocurrency exchange, you are completely anonymous, and you are the one, who controls your own private keys, instead of a trusted third party.

Additionally, I would like you to understand that once you are using a decentralized platform, there will be no service downtime, as there will be always someone who will keep up and running the network.

This is a great news, as hacking attacks also would be less likely. Because there is no central server, attacks against the network, are nearly worthless.

There are ways to slow down decentralized systems, and few theories around shutting it down, however there was not a single example until today.

Furthermore, unfortunately some exchanges only allow American citizens to trade on their websites. Moreover, the platform called Kraken, recently has announced, that they do not accept traders or investors who want to pay with British Pounds on their platform.

However if you are an existing customer, you will have a choice to change your Fiat currency to Euro, and use that for trading.

However, if you are a new customer, and having a bank account that holds British pounds, you will not be able to register on their platform.

Bitfinex, also announced, they can't have no more customers, who resides in the United States.

There are many other Cryptocurrency trading platform, that has different limitations for who can trade and who can not.

However, using decentralized trading platforms, anyone can participate, trade or invest equally.

# Chapter 12 – Your portfolio is dropping

Recently the value of Bitcoin has reached 5000 dollars, and believe me; it was party time.

However, lately there were some bad news around Bitcoin, and people seems to forgot what the Blockchain is, and how strong Bitcoin can be, and they have reacted by running away. First of all, the crypto market has been growing dramatically, especially in 2017.

There are so many people who invested only this year alone, it is fair to say that the large percentage of them does know how blockchain or Bitcoin works.

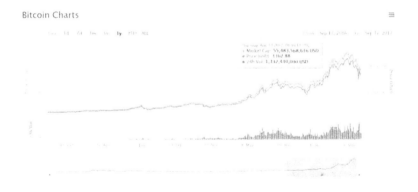

When inexperienced people see Bitcoin or any other Cryptocurrency dropping, they panic and start selling their coins. This is ok, as those who want to invest and buy even cheaper, it is an excellent opportunity. However, what you could do each time when your portfolio is failing is first, understand the reason behind the fall.

If you just start selling all your portfolio for a lower price you invested, you will lose money for sure, but you really should know what is actually happening on the market and why the drop is happening.

Next, you should look at the market, and understand what coin has fallen most, and what coin hasn't. Typically, large coins like Bitcoin doesn't drop as much as others. Bitcoin is the most stable of all currencies.

When there is a sudden drop on the market, you can see that altcoins are always affected severely. Altcoins, can take weeks or months to stabilize, however Bitcoin can stabilize within days, sometimes even in few hours.

The effect is very similar to a situation when there is a massive wave on the see. You can see the large ships will still stand like there were no waves whatsoever, while the small boats will become unstable for a longer period.

Next, understand how long the drop goes on the market. Is it a sudden drop for few hours, or is continuously going on since days. Also, you have to look at the price when it begins to increase in value again. Does it take a long time to recover? - I mean does it take longer to increase than it was dropped, or is it a slower affect? Does it look like traders have sold a significant percentage, and weakening the market?

They often do that, so new investors would panic, and begin to sell. Then, of course, traders would buy back all their bitcoins for a lot cheaper, making a good profit.

However, if this is the case, you might see that the drop and the increase look very similar to each other on the graph.

Next, you should think about how you have handled the situation, while there was a significant decline. Did you get stressed, and began to panic? If yes, maybe you should invest less.

However, if you were not stressed and you know the drop is only for a short time being, you might invest even more.

If you can handle situations, when your portfolio fallen down 20 to 30 percent, and you are comfortable with it, you are possibly a thinker for long-term success. It also means that you are more like an investor, not a speculator, and believe me, you are in a perfect condition, and you have nothing to fear.

If this is the case, possibly, you should consider investing even more, as you are certainly not a trader but an investor.

Finally, ask yourself this: Would you buy Bitcoin when is dropping again in the future?

If your answer is yes, and you are capable of sleeping at night, and not getting stressed out if you see that your portfolio is dropping, that means you are in a great circumstance to carry on investing.

I recommended you to have a plan for the next time when the value of Bitcoin will drop. One of the best strategies, is to watch the fall for a while, and see if it drops at least 20%. Once the value of Bitcoin has dropped 20%, you can buy in, however, do not spend all your money.

In case you see that the value is continuously falling another 20%, then you can go ahead and buy in once again. Since 2017 the value of Bitcoin has increased, and the market capitalization of all Cryptocurrencies have hit an all-time high.

Therefore, you will see the market recovers quickly, each time when there is a drop. As I mentioned earlier, many traditional traders began to participating with Cryptocurrency trading over traditional stocks.

Because professional traders are also presenting themselves on the crypto market, and you will see more and more often, the volatility of the value, especially with Bitcoin.

Of course, this situations will open even more opportunities for individuals on the crypto market.

However, the good news is that, individuals, now don't have to trade against traders, instead learn the techniques and methods the traditional traders use, and always think long term.

# Conclusion

Thank you for purchasing this book.

I hope the content has provided some insights into what is really behind the curtains when it comes to the future of money.

I have tried to favour every reader by avoiding technical terms on how to invest in Cryptocurrency.

However, as I mentioned few times, to fully understand how Bitcoin and most Cryptocurrency works, you may choose to read some of my other books on Blockchain as well on Bitcoin.

*Blockchain – Beginners Guide - Volume 1*

*Blockchain – Advanced Guide - Volume 2*

*Bitcoin Blueprint - Volume 1*

*Bitcoin Invest in Digital Gold- Volume 2*

The Blockchain books are focusing on the underlying technology of Bitcoin.

Blockchain Volume 1, is a beginners guide to have some basic understanding of the technology, however, Volume 2 is very technical.

Still, I did my best to use everyday English, and making sure that everyone can understand each of the technologies and their importance of Blockchain.

Bitcoin blueprint focuses on Bitcoin basics, however, I was able to add some interesting and more advanced topics around the future of payments, using reputation systems with Bitcoin-Blockchain technology.

Some of my upcoming books on Bitcoin, will provide more details on Bitcoin mining, profitability, and what exactly miners do.

Furthermore, how miners are often manipulate the market, what techniques they use, and how they try to control mining pools, using super expensive ASIC mining hardware.

I will provide information and secrets on Chinese Bitcoin miners, as well their exclusively produced ASIC mining rigs, that no other miners capable of using.

Another project that will be a continuing of this title will focus on Cryptocurrencies.

I will provide those information that I have learned on all those Cryptocurrencies that I have personally invested, and I will handpick the best Cryptocurrencies that will make you think to invest in internet money!

Lastly, if you enjoyed the book, please take some time to share your thoughts by post a review. It would be highly appreciated!

# CRYPTOCURRENCY INVESTING

*13 Most Successful Cryptocurrencies you should Invest*

# Book 4

*by*
# Keizer Söze

## Copyright

All rights reserved. No part of this book may be reproduced in any form or by any electronic, print or mechanical means, including information storage and retrieval systems, without permission in writing from the publisher.

Copyright © 2017 Keizer Söze

# Disclaimer

This Book is produced with the goal of providing information that is as accurate and reliable as possible.

Regardless, purchasing this Book can be seen as consent to the fact that both the publisher and the author of this book are in no way experts on the topics discussed within and that any recommendations or suggestions that are made herein are for entertainment purposes only.

Professionals should be consulted as needed before undertaking any of the action endorsed herein.

Under no circumstances will any legal responsibility or blame be held against the publisher for any reparation, damages, or monetary loss due to the information herein, either directly or indirectly.

This declaration is deemed fair and valid by both the American Bar Association and the Committee of Publishers Association and is legally binding throughout the United States.

The information in the following pages is broadly considered to be a truthful and accurate account of facts and as such any inattention, use or misuse of the information in question by the reader will render any resulting actions solely under their purview.

There are no scenarios in which the publisher or the original author of this work can be in any fashion deemed liable for any hardship or damages that may befall the reader or anyone else after undertaking information described herein.

Additionally, the information in the following pages is intended only for informational purposes and should thus be thought of as universal. As befitting its nature, it is presented without assurance regarding its prolonged validity or interim quality.

Trademarks that are mentioned are done without written consent and can in no way be considered an endorsement from the trademark holder.

# Introduction

Congratulations on purchasing this book, and thank you for doing so.

This book will focus strictly on the best cryptocurrencies that exist today on the market. The aim, is to learn and understand why specific cryptocurrencies are way more improved and advantageous than others. In order to identify an outstanding cryptocurrency, the best coins have been handpicked for your reference, and examined in full depth. This book will teach you, how to analyze every nitty-gritty details, and provide you with fundamental knowledge on what is most important to consider, when thinking about investing in cryptocurrency.

First, this book will begin with old school coins, those that are strongest by market capitalization, and already dominating the market. Next, it moves on, explaining enhanced technologies, such as privacy based coins. Privacy based cryptocurrencies are already dominating a large percentage of the market, and this book will show you how they diverge from one another. Next, moving on to look at the best cryptocurrencies, that are not only providing currency options, instead their primary functions are to deliver a decentralized platform. Next, you will have an opportunity to comprehend, why blockchain based platforms can change the future of doing business, by accepting the power of smart contracts, and ICO funds. The next section of the book clarifies the best blockchain-based applications, and how fiat money can be integrated interchangeably to multiple cryptocurrencies.

Finishing the book by explaining the most important facts that you must consider before investing in any cryptocurrency. This investment guide, will be 13 steps, which you must follow each time, when taking an interest in any digital currency. The bonus chapter will focus on one the most successful ICO up to date, which is a cryptocurrency, but its main function to provide a decentralized data storage platform.

There are plenty of books on this subject in the market, thanks again for choosing this one! Every effort was made to ensure the book is riddled with as much useful information as possible. Please enjoy!

# Chapter 1 – The King of cryptocurrencies

If you have been reading my previously published books, you should have a solid understanding of both technologies: Blockchain, and Bitcoin. Furthermore, my last book on Cryptocurrency Trading, I have explained how to manage a successful portfolio. Including some of my trading techniques, and strategies on what you should look out for, each time when you begin investing. Additionally, I have explained what is a wallet technology, the differences between hot wallets and cold wallets, where to buy them, as well all their pros and cons, so you can decide what best suits you. Additionally, I have also introduced the best cryptocurrency trading platforms.

When it comes to online payments, using cryptocurrencies, you must know that those transactions are irreversible; therefore I have explained the most common scamming techniques used by online thieves. Next, I have dedicated a full chapter on how to identify scammers, and how to avoid doing business with them.

Next, I have touched on Bitcoin ATM-s, how to locate them, what you need to look out for, and explained step-by-step both, how to sell or buy bitcoins, using Bitcoin ATM-s. Of course, this is an additional option to purchase bitcoins with cash only, however, people who never had bank accounts, it's not only beneficial, but might be the only way to get started in a world of Crypto.

Next, I have introduced the term: pump and dump, what it means, how to recognize it, and how to avoid losing your investment. Then I have introduced another term called: ICO. Then I have explained what ICO-s are, how to recognize the Next Bitcoin, and how to invest in early stage! Next, I covered some great strategies and techniques on how to comprehend all criteria, that must be considered before investing! This section has included how to identify your investment type!

In some cases, it is hard to tell what a particular Cryptocurrency represent: Currency, Platform, or Application, however, I have explained them all in great detail, so you should have an advantage on the market, due to this knowledge. Then, I have taken this experience to the next level, by explaining why Portfolio Adjustment most be re-visited, and provided you with examples, so you know how often you should re-balancing your portfolio. Next, I have described how to keep your wallet profitable at all times, by learning techniques on when and how to re-balance your portfolio, as well if you should outsource this task, by hiring a Cryptocurrency specialist.

Then, I have explained the fundamental differences between centralized & de-centralized Cryptocurrency trading platforms, as well their Pros & Cons that you must be aware. Finally, I have provided options on what you should do, while your portfolio is dropping. Of course, if you are an investor, and having an attitude of long-time success, you know that situations like these are nothing more, but an excellent opportunity.

Such times when the cryptocurrency market in a fall, the best times to buy and invest even more in your portfolio. However, I have never got into any specific coin yet, all though I have mentioned some by name, and provided some links to visit their platforms, in case you got interested. Still, I didn't get into any specific coins, that you should consider as some of the best cryptocurrencies to invest.

Finally, I hope you have learned what moves the market capitalization, and how to recognize the differences between market manipulation & long-term success! Once again, make sure you remember, that the value of cryptocurrency is achieved in every second by supply and demand. There are external factors, and manipulations on the market, thus traditional traders are all over the cryptocurrency market too, however supply and demand what's dictates.

Additionally, you must be aware of hackers, and unlawful scams such as pump and dump and fake ICO-s, still, it is the market capitalization that makes the value of all cryptocurrency.
This book is a continuation of my previous book called: Cryptocurrency Trading Vol 1, in case you want to recap on all my previous advice; however, I will now take an opportunity, and begin to introduce the best cryptocurrencies that you must consider, and possibly invest as soon as you have a chance.

This book will strictly focus on altcoins you should consider adding to your portfolio; therefore I will not get into much details to Bitcoin itself. In case you are not familiar with the term altcoins, it means Alternative cryptocurrencies. The reason they called alternative, is because all altcoins have been introduced to the crypto market only after the first successful cryptocurrency, called Bitcoin. Because Bitcoin became such a success, many peer-to-peer blockchain based cryptocurrency born since, and most of them are based on the grandfather of all, the first cryptocurrency, which is Bitcoin. All though, there are most of the altcoins blockchain based, it is worth to mention that it doesn't mean that all will have the same success as Bitcoin.

Because of all altcoins have been introduced after Bitcoin, it is true that most of them are somewhat copied from Bitcoin, and trying to compete against it. Smart copies of Bitcoin, have introduced better solutions, by modifying the original idea, such as: faster transaction speed or full privacy; however, Bitcoin is still the most wildly implemented in the world. All though, Bitcoin has lost a large percentage of it's total market capitalization, yet as of today Bitcoin has 47.9% dominance, against 1116 altcoins.

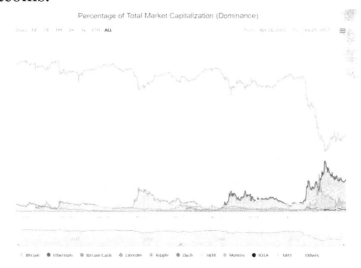

If you ask me, that's a lot of cryptocurrency against Bitcoin. The reality is, that most of them are not trying to compete against Bitcoin; instead, they provide a different solutions for a different problems.

Some of them are offering platforms for specific applications, and some are the applications themselves. And of course, most of the cryptocurrencies, probably 80 to 85% of them are entirely useless, and in the next few years, they will disappear forever.

In the same time, there will be many new currencies born, and try to complete not only Bitcoin, but other altcoins, providing better solutions for cheaper or faster transactions, more decentralized systems, or improved proof of steak manner. Either way, my number one investment for you is Bitcoin. I have written two books on Bitcoin and explained everything there is to know about it.

*Bitcoin Blueprint - Volume 1*
*Bitcoin - Invest in Digital Gold- Volume 2*

Additionally, I have a bundle book that contains both books for a special discount which fully explains the birth of Bitcoin, and it's timeline, up until today.

## Bitcoin: Guide book for beginners – 2 books in 1 - includes: Vol1 + Vol2

In case you are new to Bitcoin, and thinking that Bitcoin is so expensive, and don't have enough money to buy a whole Bitcoin, then let me explain something quickly. You don't have to buy an entire Bitcoin. Instead, you can start with 100 dollars worth of Satoshi, which is around 0.02659 BTC. – Bitcoin.

There are 100,000,000 satoshis that equal to 1BTC; therefore there is enough for everyone in the whole world to get some. Satoshis are named after the creator of Bitcoin, and you can look at them as a change value, similar to cents with dollars, or pence with the British pounds, satoshis are the change for bitcoins.

As of September 2017, the current circulating supply of Bitcoin is 16,583,075BTC, where the maximum quantity that will ever be on the market, is 21,000,000 BTC by the year of approximately 2140.

Again, I will not cover everything in this book about Bitcoin; however, you must understand that it is the most valuable currency that ever existed. It has taken over all other traditional Fiat currencies within few years, and even hundreds of other enhanced cryptocurrencies not near valuable as Bitcoin, therefore if you wish to get into the cryptocurrency market, the number one investment you must choose is Bitcoin.

In case others tell you otherwise, and you are confident that other currencies can be just as profitable as Bitcoin, then, by all means, do what you wish, however, personally I don't trust in any cryptocurrency as much as I do in Bitcoin.

Bitcoin has been surviving all those bad reputations that it had to deal with, including accusations of parallel uses on the deep web for illegal activities, as well all those hacking attacks that trading platforms have suffered. Hacking attacks were especially decreased the price of Bitcoin, because it suggested people to stay away from it.

Then again, what really happened is people haven't kept their values secure enough, and it was open for hackers, and of course they jumped right at it.

Then, Bitcoin banning all over the world, now it's all in a process of change, and pretty much in every significant country, Bitcoin is legalized. Even though there are some recently announced news about Bitcoin to be banned in China, hasn't made any real fall for Bitcoin; instead, it seems that more investors began to jump in. Additionally, China prohibits Bitcoin? Common' now!

China has banned Bitcoin in the past, then suddenly it's all been forgotten, furthermore, if you analyze the trend and what China did in the past, you will quickly realize that it's only another performance act, nothing more. China has banned Google, as well Facebook in the past; however, those two companies just became even stronger, and more valuable.

The list of China was banning this and that and the other, could go on and on; yet, you have to know something more important.

The large percentage of Bitcoin mining is happening in China. We are talking about close to 65 to 80 % of all Bitcoin that was ever mined, happened in China.

This is mainly because the electricity is one of the cheapest there, however, when it comes to mining, you also need super expensive mining equipment. This super sturdy material's called ASIC Bitcoin mining hardware.

Yes, you guessed it! They are made in China, and there are even more specialized hardware that is not for sale; instead, it's kept for themselves, so they can have access to more profitability than anyone else on the world.

You think because you have money and you can just buy the best ASIC mining hardware on the market, however, those are not for sale, or at least for anyone around the world. Again, the Chinese are not in the habit of making mistakes or exhibiting dumb moves, especially nothing that could hurt them financially.

The technique they use instead, is to create fear on the market, making everyone stay away from something that they can later purchase cheaply, and take control over it. Eventually, sell that product ( in this case: bitcoins) on the highest reasonable price, while they will be still the cheapest on the market to buy from.

Anyhow, I didn't intend to go as deep, when it comes to Bitcoin, as I already wrote two books on this topic. I already explained facts, why Bitcoin is here to stay, and many countries looking at it as changing it over the dollar. Some might think, it's impossible because of its volatility; however, you have to understand that even Fiat currency is volatile.

In fact, an essential Fiat Currencies such as USD or GBP, are so unpredictable, that sometimes loses 10 to 20 % of its value within overnight.

What you have to understand is that Bitcoin is only used by less than 1% of the world population yet, and once there will be an increase in its usage, it's volatility will decrease. Of course, by then the value will also increase dramatically, but as I mentioned twice already, here it comes my third one:

**My number one recommendation is to invest in Bitcoin!**

## Chapter 2 – Litecoin

First of all, I would like to mention that the first cross chain atomic swap has been completed between Litecoin and Bitcoin. The first exchange between two different blockchain platform, has been done by Charlie Lee, the creator of Litecoin. He is also a Director of Engineering at Coinbase, and recently has announced that he was able to swap 10Liteoin to 0.1137 BTC.

Cross chain atomic swap is a way to trade cryptocurrencies between each other, without a trusted third party to be involved. Due to its cross chain method, it will be possible that we all can use in the future, to swap cryptocurrencies between ourselves, having no trusted third parties to involve.

As of now, Litecoin seems to be the center of all other currencies thanks to Charlie. The proof of concept was published in 2013, however many people were skeptical about it, but due to Charlie's demonstration of a successful swap, it will make some dramatical changes of the way we do business.

Additionally to the good news, when analyzing the recent history of Litecoin for the last year, you can see a huge increase.

Over a year ago the value of Litecoin was about 3 dollars, however recently it has reached 90 dollars, meaning immense profitability. Using dollar as an example, if you were to invest $100 to Litecoin a year ago, it would worth today 3000 dollars.

Now, that all those good news are coming out, I wouldn't be surprised if Litecoin would take off even further. Additionally, you can see there is no pump and dump on the market, as it's going one way in value, which is up.

Now that's being said; it's nice and groovy; however, you may ask the question: What on earth is Litecoin? Well, let me tell you a little about it. Litecoin was created back in 2011, which makes it one of the first altcoins in circulation. As I mentioned earlier, Litecoin was produced by Charlie Lee, and for the primary purpose of its solution, is to answer all those issues that Bitcoin was facing.

First, Charlie has reduced the block time creation, that is around 10 minutes with Bitcoin, with Litecoin is about 2.5 minutes. Primarily every transaction gets validated on the blockchain once there is a new block sealed. Because in the case of Litecoin each block gets validated in every 2.5 minutes, all transaction becomes validated four times faster on the Litecoin's blockchain.

Additionally, to reducing transaction validation, Charlie was also able to reduce the transaction fees. Litecoin was also the first cryptocurrency that was using the script algorithm, which is a bit different than Bitcoin's algorithm SHA 256. However, Litecoin is nothing more, but a modification of Bitcoin, as all the mechanism that was created with Bitcoin, followed through Litecoin. What you have to understand is that Litecoin, as far as the development goes, it's always been a step ahead of Bitcoin, and the reasons are simple. Because Bitcoin has such a vast community, when issues arising, everyone is scared when it comes to making changes in the network. Each revision on Bitcoin's blockchain must be approved by at least 75 – 81% of all miners, and in a case of Bitcoin, it takes a long time to achieve such results. However, because Litecoin has a smaller community, and less risk when it comes to possible losses, Litecoin miners are more flexible, and changes get approved lot faster than in a case of Bitcoin.

Because Litecoin is capable of evolving newer technologies then Bitcoin, it is highly likely that Bitcoin will continue to improve, but Litecoin will continue to grow at a faster rate. For that reason, Litecoin seems to be the solution provider, by providing the newest and greatest technologies on the blockchain first, before Bitcoin.

Each time, it seems that Bitcoin is waiting for Litecoin to test specific changes, then Bitcoin adopts it if they wish to do so. However, in some cases, this could mean that Litecoin can be ahead it's game then Bitcoin. Litecoin has a current market capitalization of 2.6 Billion dollars, making it the 6th biggest cryptocurrency of all. Either way, when you compare the historical charts of Litecoin to Bitcoin, they both look alike, all the time. Additionally, when you are looking at the competition of Litecoin to Dash or Monero, (more on these later) the volume of Litecoin comes out on top. The volume tell you the amount of people are using the currency, which might differ from the market capitalization, as well it's value. When you see large volume, all it means is that people are interested in using the currency, which is defined te amount of transactions happening on the blockchain.

Another good indicator is the transaction fee. When it comes to transaction fees, everyone wants to pay less as possible. This is no science; however, Litecoin has a considerable advantage when it came to transaction fees, especially when it compared to Monero or Dash.

| # | Name | Symbol | Market Cap | Price | Circulating Supply | Volume (24h) | % 1h | % 24h | % 7d |
|---|---|---|---|---|---|---|---|---|---|
| 1 | Bitcoin | BTC | $62,847,672,818 | | | | | | |
| 2 | Ethereum | ETH | $28,829,817,101 | | | | | | |
| 3 | Bitcoin Cash | BCH | $7,071,640,519 | | | | | | |
| 4 | Ripple | XRP | $6,822,136,348 | | | | | | |
| 5 | Dash | DASH | $2,666,763,622 | | | | | | |
| 6 | Litecoin | LTC | $2,615,788,492 | | | | | | |
| 7 | NEM | XEM | $1,996,660,000 | | | | | | |
| 8 | IOTA | MIOTA | $1,466,640,237 | | | | | | |
| 9 | Monero | XMR | $1,289,113,578 | | | | | | |
| 10 | Ethereum Classic | ETC | $1,002,500,661 | | | | | | |

The price prediction for Litecoin, will follow closely to Bitcoin. As long as Bitcoin keeps performing well, Litecoin will equally play well as well. Litecoin, indeed has a potential to increase even more than Bitcoin, mainly because of the lower transaction fees, could adopt additional customers, then Bitcoin.

Litecoin in value, could increase more than Bitcoin's value, not only because it has more potentials, but Litecoin will continue to implement technologies faster, the growth potential for Litecoin could far exceed the growth potential of Bitcoin. Since January 2017, up until now, Bitcoin growth has been 4-5 times more significant; however, Litecoin has increased in its value 25 – 30 times.

This means that with a small amount of money such as only 100 dollars, Bitcoin was probably generated an additional 400 dollars in less than a year; however, Litecoin has created over seven times more than Bitcoin. Litecoin also has a limited supply as Bitcoin, yet in a case of Litecoin is 84 million LTC. Currently, the mining award for each block validation, is 25 LTC, that will respond similarly to Bitcoin, which will halves in every four years until it reaches of a total of 84 million Litecoins. For further information and visit Litecoin website follow the link provided:

### https://litecoin.org/

## Chapter 3 – Privacy based coins

One of the major topics that are all over the headlines in the crypto world is nothing other than privacy. The main reason that Bitcoin got adopted in the first place is indeed privacy. Cryptocurrencies wouldn't be where they are today without that little sense of anonymity, and privacy.

Unfortunately, there are more and more is seen, that Bitcoin and it's leading competitors are not providing full privacy. This is because all the transactions are visible on the blockchain. Of course, this issue have been causing fear for a while, and lots of people began to search other alternatives then Bitcoin.

When you think back in time, the way we used to transact, cash was one of the best solutions to privacy. I mean, there are other ways to negotiate amongst us, using other alternatives such as Checks, Wire transfers, Credit Cards, Swift, or other trusted third parties such as PayPal, Payoneer; however, none of these mechanisms are providing anonymity. Don't get me wrong, as I am not saying that these solutions didn't help us, in fact they still exist and probably won't go away as smoothly as we might want to. However, these types of solutions are decreasing our privacy. You might don't have any problem sharing with the world where you spend your money, yet a large percentage of people do prefer privacy. Either way, cash, I am referring to paper money is one the best model to look at, especially the way it works in real life.

Money exchange in real life, is what really should be enhanced for the internet. Here is what I mean; if I give you a 10 dollar note, then you give it to your friend, there is no direct link where that money came from in the first place. For example, the bank can tell how much money I have on my bank account; however, the bank can not know precisely how much cash I have in my pocket. Yes, paper money has serial numbers and somewhat can be traced, however, the amount of money I have, only can be revealed by me. Either way, cash is regulated by the government, and it can be easily counterfeited; therefore cryptocurrencies can become an excellent choice to replace paper money.

Privacy is not all about criminals, just because there are ordinary people also prefer privacy, instead of publicity. For example, many people prefer not to advertise their spending habits. Personally, I don't want the world know what I just bought on another website last week.

Additionally to Bitcoin, it is somewhat anonymous yes, however, if you give me your wallet address for the soul purpose of making a payment to you, it will be validated on the blockchain. The problem here, is blockchain not only recording the transactins I have initiated to you, but all purchases that you and I have ever made. We will see each other's funds, which we have control over with those wallet addresses that is yours and mine.

This website shows you all live Bitcoin transactions taking place:

### https://blockexplorer.com/

You might be a nice guy, however, if I choose to pay someone with Bitcoin, they can easily see all my transactions, and my wallet can become a target. Business relationships are also something that I should mention here. Currently, we can take an example of a bank, however, in the future, we can look at a decentralized trading platform, where I would want to keep all my funds.

Once again, my wallet can be visible to any hacker, and my account can become a target. The reality is that less than 10% of the people cares about privacy when it comes to online shopping. However, with cryptocurrency we are moving to a different level. That 90 % of people who don't care much about online privacy, they say that because it's all about online shopping.

However using fiat currencies, by paying someone, the person you are paying, cannot reveal how much money you have on your account. Neither you can see how much she has on her account. But again, in a case of Bitcoin, it's all on the blockchain.

There are few great privacy coins out there which are now solving these issues. Some run on the blockchain, some uses other technologies instead; however, I will now reveal the most well-known privacy coins, and you can decide which one you prefer best.

# Chapter 4 – Monero

First and foremost, Monero is the most anonymous cryptocurrency currently exists. There are many other cryptocurrencies advertises itself that is anonymous, and somewhat it is true, however, when it comes to Monero, we are talking about an entirely different level. As I mentioned before, all cryptocurrencies are anonymous, as there are no names mentioned on the blockchain; however, the wallet address is visible, as well all transactions that were sent or received from that wallet. Monero has been designed, somewhat in a different way.

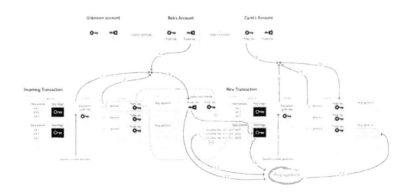

One of the reasons that Monero is more anonymous than other privacy coins is because, Monero automatically creating dummy addresses, and uses those to send the funds through to its original destination.

For example with Bitcoin, I can send you 100 dollars, initiated from my wallet to your wallet, then once it's validated, it will be recorded on the blockchain forever.

But as you see all there are two wallet addresses participating in one transaction, nothing more. However, Monero in the other hand works differently, and the best way for you to quickly understand how it works is to take a look at an example.

First, let's imagine that I am about to send you $100 worth of Monero. Instead of one transaction, there will be multiple transactions. Typically between 3 to 6 sales that each transaction would represent. Some part of the founds I am about to send you over would use multiple addresses on it's way of the transfer.

It is not visible, however, let's imagine that my 100 dollars worth of Monero transaction, will create five transactions where 3 of transactions will represent 20 dollars worth of Monero, one transaction will constitute of 30 dollars, and another transaction will constitute another 10 dollars worth of Monero. What happens it that one of those five transactions will reach another wallet, that was never existed before, which in this case I call Jack's wallet. What happens next, is Jack's wallet is controlling many other dummy wallets, that Jack's wallet will use to send you the payment.

Now, let's imagine that Jack's wallet has only used the first 20 dollars worth of Monero. Then as you see in the same manner, there will be another four transactions where each of the transactions will be forwarded to another wallet that controls hundreds if not thousands of other wallets, and each will be used to transmit the money to you.

However, when you receive all those transactions, it will seem like it was only coming from one wallet as only one transaction. This method is known as a ring signature algorithm.

Again, all those transactions are not visible to anyone, and automatically created by the system. Some of those addresses that Monero uses, either brand new, or not been used for a while to transport any Monero.

As you see Monero's blocks are not providing the addresses, instead only the hashes, fees, and sizes:

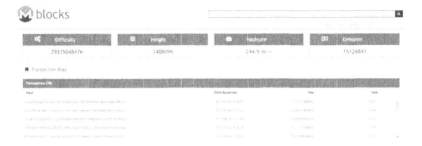

Additionally, Monero has no scaling problems such as Bitcoin has, and it's also using proof of stake; therefore Monero is minable. Also, Monero is a currency that anything can be bought, with complete anonymity; consequently it's been used very heavily on the Dark Web.

Monero currently is the 9th place having 1.3 Billion dollars worth of Market Capitalization, and now the value of Monero is around 88 dollars.

However over the last year, Monero's volume has been significantly increased, and the amount has been moved from 5 dollars to it's recent highest of 144 dollars.

Even though, right now there is a market crash, still, Monero holds itself around 88 dollars, which is around 16 times more than a year ago.

Overall, Monero provides confidentiality to both the sender as well the receiver, and nevertheless the amount that has been sent over the network.

To visit Monero's website and find out more information, you may use the link below:

**https://getmonero.org/**

## Chapter 5 – DASH

Dash was developed back in 2014 by Even Duffield, however back then it used to be called Darkcoin, which has been changed to Dash - Digital Cash in 2015. Because it has a Bitcoin code base, Dash is compatible with all existing merchants, exchanges, and wallet software that has been previously was used with Bitcoin.

Dash is also a privacy coin, and it's based on the blockchain Bitcoin technology; however, there are few differences. First of all, Dash has larger blocks then Bitcoin blockchain has, meaning more transactions can be fitted to each block, which results in lower transaction fees. Dash also uses master nodes. Bitcoin network also uses master nodes, however, in a case of Dash, there are some additional features. The master nodes on the Dash network, each has a deposit of 1000 dash on them. Those Dash coins are not meant to be moved from the master nodes;

otherwise, the master node would stop their function. These master nodes are responsible for two additional features on the Dash network. One of them called InstantSend and the other element called PrivateSend.

InstantSend means precisely what the name implies, which is approving transactions instantly by the master node on the network, therefore the transactions can be confirmed immediately. This is great, as you don't have to wait for an additional six transaction confirmations to be completed anymore.

PrivateSend, once again, it's what the name implies too, which is allowing you to have privacy when you are sending your Dash coin on the network. However, there are some limitations, which are expect you to have a set limit of a minimum amount of Dash to be transferred to the system.

The denominations are 0.1 Dash, 1Dash, 10Dash and so on. The other issue is that these transfers can take as long as between 6 hours to 3 days to be completed. However, these transactions will be completely anonymous.

Dash block time is in every 2.5 minutes, and its mining difficulty can adjust up to every single block using an algorithm called Darg Gravity

Wave. Dash current block reward is 3.88 Dash, and this amount is decreased by 7.1% in every year. The total Dash, once all mined, will be 22 million Dash coins. Dash block reward is paid out to three separate groups:

• 45% to miners for security
• 45% to the master nodes, including InstantSend & PrivateSend functions
• 10% is the treasury to be paid out to developers, marketers, or anyone else who is approved for a project.

Dash is a 5th largest Cryptocurrency with a total market capitalization of 2.5 Billion dollars. The reality is that Dash seems to have more to offer then Bitcoin, however Bitcoin is the oldest cryptocurrency; therefore, Dash will have a pretty hard time to compete against Bitcoin. Still, Dash confirmation time is lot's faster then Bitcoin, also provides more anonymity, and the transaction fees are a lot cheaper than in a case of Bitcoin.

Now, there is one thing that you must know, which is Dash provides privacy too, however, becase it's a blockchain technology, and not cryptonote as Monero, I still believe that Monero could give much better privacy. Once again, Monero did have some issues previously, due to Monero is not precisely blockchain, those vulnerabilities are now fixed.

## Dash Marketing

Dash has been capable of doing some excellent Marketing, and by doing so, their community has been grown dramatically.

Still, some people might not believe that marketing is useful, the fact is that you might create the best cryptocurrency ever, yet, if you don't promote it, and don't advertise it well, people will never going to use it, merely because they are not aware of its existents.

Marketing is what helps to take these cryptocurrencies to those who should use it; therefore I genuinely believe that Dash has been doing some excellent work so far. This is why, I also highly recommend to check out Dash, think about investing, as even Bitcoin is on the top of the market, you never know what the future can bring us.

Dash has been doing pretty well since last year, it has grown more than 35 times in value. In 2016 the amount of Dash was about 10 dollars, however recently Dash has reached its all-time high cost of 395 dollars.

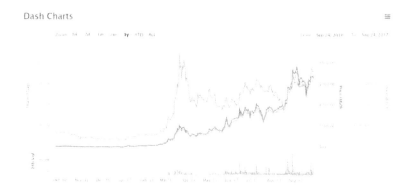

To find out more about Dash, and where they are heading to, you can use the following link to their official website.

**https://www.dash.org/**

## Chapter 6 – PIVX

Firs, let me begin with the fact that PIVX is also privacy based coin, which stands for Private Instant Verified Transaction. PIVX first was called Dark Net; however, it has been rebranded in January 2017.

PIVX block time is 60 seconds, and the total supply that ever be on the market is 43 million PIVX coin. PIVX has a really nice community, which is always very good to have supporters, thus for the coin to take off, you will need lot's of followers, and a many admires, which PIVX has. Additionally to their community, PIVX has an excellent looking website, and their marketing strategy is also unique.

PIVX has multiple forums, YouTube channel, Twitter account, however more importantly a constructive community. This is vital, as in case you have any questions, you can get a quick reply, which not only helpful but always makes sense in terms of the quality reply.

Some of those cryptocurrencies where you don't get a reply for your question, or the answer is not clear; I highly recommend to stay away from, however, when it comes to PIVX, it's whole different story. I mean, who would want to invest in a company where there is no communication right?

The algorithm that PIVX uses is called: Quark. Additionally to Quark algorithm, PIVX applies proof of steak 2.0

PIVX has been build up as a fork of Dash, with a little mixture of Bitcoin Core; therefore it's a fascinating hybrid technology. PIVX has started off back in January 2017, when they also created their wallet software, however, since they have been building many more exciting features.

Such features are called Multi Signature Access Control, and a new Multi Sig Escrow system. PIVX also offers SWIFT Transactions, which confirms transactions in seconds, that's guaranteed by master nodes without having to wait for multiple confirmations for validity. Basically all it means that PIVX is a lot faster than Bitcoin.

Another great feature is they have managed to have a built-in Encrypted Chat system to their wallets. Again, this is something that no other cryptocurrency company has ever achived, especially within privacy based coins. PIVX is one of the most significant players in the privacy industry within all cryptocurrencies.

To my mind, Monero is the best privacy coin that currently exists, and I am not sure if I should believe that PIVX can reach the same value as Monero.

However, because there is an urgency on the market regards to privacy based coins, it is fair to say that PIVX will increase in value over an extended period. PIVX still a reasonably new currency on the Crypto market, however, it's currently stands on the 35th place with a total market capitalization of 157 million dollars.

The value of PIVX, back in January 2017 was 0.007 dollars, however, recently has reached its all-time high value of 4.45 dollars. This might don't seem a lot to you, yet within seven months, the value of PIVX has been increased 635 times.

This mean that if you would invested $100 into PIVX, back in January 2017, today it could worth around 63,500 dollars. Being honest here, there is no guarantee on where Civic will stand within a year, however, due to its nature, and supporting platform, it can quickly go to the same directions like Dash or Monero.

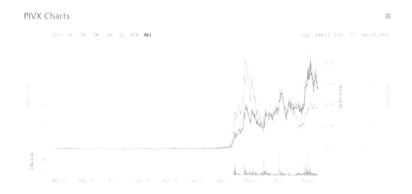

For further information, and additional technical analysis, you may go ahead and visit their website at the following link.

## https://pivx.org/

# Chapter 7 – Zcash

Zcash is another privacy based cryptocurrency, which is also open-sourced, and completely decentralized. Zcash is having strong privacy protections, amonst many otherfeatures, so let's begin to learn about it.

The Zcash network has been launched back in October 2016. Once again, Zcash is based on Bitcoin. Zcash transactions are shielded by hiding the recipient, the sender, and the value on the blockchain. Zcash encrypts the contents of shielded transactions.

With the encrypted information, the protocol uses a cryptographic method to verify the validity. It uses a protocol called: zero-knowledge Sussinct Non-interactive Arguments of Knowledge, which is also known as zkSNARKs. This contract approves Zcash network to maintain the secured balance on the ledger, without revealing the parties or the amounts involved in the transactions.

All metadata is encrypted, and used to prove, that nobody is cheating on the system. zkSNARKs is what makes the Zcash so private, and zero knowledge proves the Zcash's token feature. Transactions are less than 1Kb in size, and take less than six milliseconds to verify.

Although Zcash mainly knows as a privacy-based coin, still it is possible to use it publically if it's chosen that way by the sender. However, a payment that sent from a shielded address to a transparent address will reveal the received balance, while payment from an open address to a shielded address will protect the accepted value of Zcash.

The founder and CEO of Zcash are Zooko Wilcox, who has more than 20 years of experience in decentralized systems, cryptography, information security, and start-ups. Additionally, the team behind Zooko, is working extremely hard, and keep on creating new features with Zcash. Some of the top advisors within the group are Gavin Andersen and Vitalic Buterin. Gavin was one of the first people who got involved in Bitcoin, and currently he is a Chief Scientist at the Bitcoin Foundation. Vitalic is a founder and Chief Scientist of Ethereum.

There are also many other well known investors, and it's worth to mention that one of the leading people is Roger Ver, formerly known as Bitcoin Jesus.

Zcash is currently sitting as a 16th biggest cryptocurrency, with a total market capitalization of 517 million dollars.

| 10 | NEO | NEO | $1,224,675,000 | $24.49 | 50,000,000 * | $29,487,300 | 2.87% | 24.40% | 18.87% |
| 11 | Ethereum Classic | ETC | $1,036,777,749 | $10.82 | 99,869,376 | $33,297,100 | 0.47% | 3.34% | -6.04% |
| 12 | OmiseGO | OMG | $983,916,567 | $10.01 | ** *** *** * | $58,804,400 | 0.11% | 15.63% | -5.28% |
| 13 | BitConnect | BCC | $796,504,350 | $117.49 | 6,779,232 | $10,249,500 | 3.18% | 6.32% | -9.83% |
| 14 | Lisk | LSK | $691,901,206 | $8.14 | 112,715,000 * | $3,340,620 | 0.53% | 0.93% | -5.61% |
| 15 | Qtum | QTUM | $594,466,220 | $8.00 | ** *** *** * | $72,370,600 | 5.81% | 11.40% | -3.96% |
| 16 | Zcash | ZEC | $517,957,271 | $229.01 | 2,261,604 | $33,484,300 | 5.60% | 15.40% | 20.67% |
| 17 | Tether | USDT | $442,870,821 | $1.00 | 442,481,207 * | $148,619,000 | -0.15% | -0.26% | -0.27% |
| 18 | Waves | WAVES | $422,863,000 | $4.23 | 100,000,000 * | $5,773,120 | 0.47% | 0.66% | 0.62% |
| 19 | Stratis | STRAT | $416,125,415 | $4.22 | 98,567,298 * | $4,756,630 | -5.28% | 3.11% | -16.87% |
| 20 | Kyber Network | KNC | $322,526,078 | $2.34 | 137,895,000 * | $18,887,300 | -1.37% | 10.20% | ? |

The current value of Zcash is currently 228 dollars, however, in June 2017, the value of Zcash has hit an all-time high of 410 dollars.

You might ask the question: Is Zcash better than Dash? Or Monero, or PIVX?

Well, the reality is that I truly believe in Monero; however, even Monero has it's pros and cons; therefore It is difficult to say which privacy coins will win the game. I genuinely believe that all these privacy based coins I am introducing to you will stay on the market for a very long time.

Will there be any winner? The thuth is that none of them are perfect, in fact, there is no ideal cryptocurrency out there.

It doesn't matter which one you choose, they will all have many issues in the long run, and it will entirely depend on the team behind of each coin, as well what kind of attitude they will have when it comes to continuous development.

I suggest once again, that it's not only the features that need to be looked at, but also the development team, and how determent they are to continue to develop their technology.

This brings me to my point, which is this. If you see people behind the project such as Gavin Andersen, Vitalic Buterin, or Roger Ver, you have to understand that their reputation has been in light for long years already, and it's highly unlikely that they will leave their projects behind.

For further information on Zcash, you may go ahead and visit their website on the following link:

**https://z.cash/**

## Chapter 8 – Ethereum

Ethereum was proposed in 2013 by Vitalic Buterin, a Russian-Canadian programmer. Vitalic was 19 years old at the time. However only in 2014 was capable of receiving enough funds for his project, which went live only in July 2015.

Ethereum also an open source distributed blockchain- based system, however, it's real power is to provide a distributed computing platform, instead of a payment system. All though Ethereum has its currency called Ether, Ethereum's primary purpose, is the platform itself.

While I have only explained Cryptocurrencies that's primary purpose is to make a payment, Ethereum is a platform. Most people believe in the technology and how it can change the future, like the internet. Still, many people just want to jump in, and make a quick buck by investing in Ethereum.

No wonder, of course, the value of Ethereum has been grown more than 3000 times as it was a year ago. But, what Ethereum is? Well, Ethereum can represent multiple services simultaneously. The best and the easiest way for you to understand is if I explain some of its core features.

**Currency**

One of Ethereum's feature is a digital currency; therefore it can act in the same way as Bitcoin. As I mentioned earlier, Ethereum is blockchain based; however, Ethereum has 15 seconds block times.

While Bitcoin has about in every 10 minutes. Ethereum is validating transactions on the blockchain in approximately every 15 seconds.

Right there, you probably feel the power of Ethereum already right? I can't blame you, but let me tell you, that this is just a tip of the iceberg.

## Smart contracts

Ethereum can create smart contracts, but first of all, what are smart contracts? Well let's begin by explaining you that smart contracts are pieces of codes that live on the blockchain. These codes can execute transactions, without human involvements.

The way smart contract works, in case you want to create your smart contract, first you have to download a specific browser called: Ethereum mist. This browser can interact with Ethereum DAPS (Decentralized Application), this is needed so you can access the smart contracts functionality.

The Ethereum mist browser comes with a complete copy of Ethereum blockchain. Once you have that, you have to navigate using a special browser to reach the DAPS browser to find the smart contract, that is hosted within the DAPS. Here, you would see the smart contract that has been written for you by Ethereum developers, and once it's ready for you to use it, you can enter the parameters that you want to achieve.

Sounds difficult already, however, there are some other things you should be aware of too, which is cost. You have to pay Ethereum transactions fees, pay a fee to the smart contract company, and pay the general charge of any inflationary cryptocurrency, which is consistently creating new coins that makes its infrastructure larger.

Right, so, such platforms like smart contracts for anyone? I don't think so.

How about Cryptocurrency that people can use to make payments in exchange for goods, is that for anyone? More likely.

Business applications are essential for medium-size to large-size businesses, maybe even small companies, however, it's not for everyone. Should you consider to invest? Yes, you indeed should, as a platform like Ethereum can change the future of doing business.

Contracts will always be required for people, even if you just rent a car or a flat You probably have a contract with your Internet Service provider, or when you are purchasing a property, once again, contracts are continuously exchanging hands.

However, in my personal view, smart contracts are for those companies, who uses such technology in their daily life. I think, that individuals don't need as much of a smart contracts yet, and this is simply because most people don't understand how they can be useful. The theory here that is also representing the big boom is once again: blockchain, but this time the payment system isn't the main purpose, instead the smart contracts are.

Of course, payments might involve in the process, especially if you want to purchase a property, so let's take an example here. Imagine that you are about to buy a property. With the current process involved, you have to go and see a Real Estate Agent who will show you the flat or house, and once viewed it and ready to make an offer, the current owner should accept your offer.

However, cases like that buyers never interact with the sellers, instead they are using the Estate Agency. That's another long story, however once that's complete, the paperwork began to take place. This will consist of all issues with the property, including existing debts, or previous modifications and so on.

All these documents will be provided by the owner, who will pass these reports to their solicitor, who then validate them all. Once it's approved, the current owner's solicitor will give those documents to your solicitor, who will double check everything and make sure all the documents are validated correctly, then will be given to you. Once you view all those records, you might go ahead and make a payment, or call your bank to validate a mortgage to the specific property in question.

I will simplify this process, by saying that you are ready to make a payment without bank or mortgage because maybe you are buying from a friend or from someone who has a better reputation.

So let's imagine that the property details are already on the blockchain and previously validated by the blockchain, and all you have to do is create a smart contract and make a payment for the property online.

Of course, when you would do that, it would also mean you could terminate the need of Real Estate Agents, as well all the solicitors, as you wouldn't need their approval. This is because all reports would be already validated on the blockchain.

Again, using smart contracts can change, and will change the way of doing business. Ethereum's smart contracts has it places when it comes to a blockchain platform. Awesome, however I have bought two properties already, and when it comes to validating documents, I will be completely honest with you and tell you the truth. Each time, I hired Mortgage brokers and solicitors who can doublecheck and warn me what to look out for, or at least I should be aware.

Let's say that the property is already on the blockchain, and all its details are validated, however still I would rather hire someone to doublecheck everything making sure, all is indeed as it should be. This is because I am not a Real Estate Agent, neither a solicitor and most people are still sceptical when it comes to investing in real estate.

Therefore, when it comes to sell and buy property, once again I will hire someone again, but that's just me, and there are billions of people who think otherwise.

Anyhow, Ethereum has it's place, and maybe I came up with the worse examples, but the truth that there is a huge market for smart contracts, which will only going to increase in the future.

## DAPS

Decentralized applications. Now we are talking. These applications are run on Ethereum and believe me; this industry is already huge. DAPS can be used to build online gaming platforms, mobile applications, or social media platforms like Facebook or Twitter. Instead of having thousands of servers that you have to pay for hosting, you may build it over Ethereum. This is another critical feature of Ethereum, and it's also unique.

**Crowdfunding platform**

In volume 1 – Cryptocurrency Trading, I have explained what an ICO is, however, let me provide a brief description, in case you are not familiar with the terminology.

ICO stands for Initial Coin Offering. It is similar to the Initial public offering with companies who are offering shares from a company in exchange for an investment; however, ICO-s are offering some of their coins for the exchange of investment.

Once a company comes up with a new cryptocurrency, they can receive investment in Ether, if they will host the newly created coins on the Ethereum blockchain.

The reality is that most of the new cryptocurrency companies are entertaining their new idea on Ethereum blockchain, and their funds are received in Ether. Once again, the industry is enormous, and it doesn't seem to go away; instead, it will expand even beyond, in the future.

As you see there are many potentials with Ethereum, however once again, the value of Ethereum is made by supply and demand. There are some serious tech giant companies already invested in Ethereum, which increased the price in value; however, when that happened there were some side effects that you should be aware.

Many people just want to make a quick buck, and when the value of Ethereum was on the rise, many people who were entirely new to the crypto market has invested in Ethereum, not knowing even what Ethereum is representing in the first place.

This side effect has increased the price again, which has created another domino effect, and those who realized that the price is on the rise, quickly jumped on board too, simply because of fear of missing out. FOMO. This term also used on the crypto market very often, simpy because lot's of people only investing because they FOMO.

# Ethereum VS Ethereum Classic

If you are getting confused, don't worry, shortly you will understand what's going on. There are two currencies of Ethereum, but how did this happen? Well, let's go back in a year, and let me explain all about a failed projects first.

The company called DAO, Decentralized Autonomous Organization was building its project on Ethereum; however, the company got hacked when they have announced their crowdfunding. Funds were stolen from investors, which at the time was worth more then $50 million dollars.

Now you might think that isn't a lot of money, however back then was only one kind of Ethereum, and the majority of this one currency has been stolen by one transaction. This, of course, was going to cripple Ethereum as a whole, and its value could have been quickly become worthless.

What happened is the Ethereum foundation as a whole had to discuss what they are going to do about it. But before we move on, first make sure we understand the fundamental principles when it comes to blockchain transactions. Once, a transaction has been completed, it is irreversible; therefore there is no refund or anything like that. All transactions that were once made are done, and in the Ethereum blockchain platform it's not even 10 minutes, but as I mentioned earlier, the block times are on average of 15 seconds. Anyhow, back to Ethereum foundation.

They held a meeting and asked for a vote in regards to the theft. They have 80% of the Ethereum creators who stayed with the option of reverting that one transaction, and 20% of them against it.

This has resulted in a hard fork, meaning the chain on the blocks has divided into two separate chains. The one with the 80% of voting stayed with the name of Ethereum, and the other chain that voted by the 20% of Ethereum community is known as Ethereum Classic.

**Fake news – does it matter?**

The forum called: http://boards.4chan.org/biz/ where a post came through on the 25th of June this year, which was something like that:

,,*Vitalic Buterin is dead. It's now confirmed ,there was a fatal car crash."*

This post, wasn't even confirmed by anyone, still it seems to be responsible of wiping out 4Billion dollars worth of Ethereum in less than two hours. At one point Ethereum value went down to 10 cents. The recovery was fairly quick, as once Vitalic heard the internet rumour, he quickly took a selfie, and posted the latest block number on Twitter. However, Ethereum has lost around 10% of its value that day.

Source is from Vitalic's twitter account:

**https://twitter.com/VitalikButerin/status/879127496024772610/photo/1**

### Where do we stand today?

The fact remains, which is a massive demand for Ethereum, especially internationally. China, Singapore, and Russia of course, as well many other countries who continuously contributing to Ehtereum, therefore it has a great future.

Ethereum has so many projects that I could write multiple books about them; however, I wanted to provide some overview by looking at not only the upside but cons too.

What you have to learn from Ethereum, is that even a small fake news can manipulate the market, so severely, that it's very difficult to predict what will happen tomorrow.

In the same time, you have to understand the recovery is speedy, which means that even there are some issues; still, there is a vast interest in the technology.

**Summary of Ethereum**

In summary, you should consider taking an interest in Ethereum, however, if you are looking for a quick buck, and want to make a 10% or even 20% profit every day, I wouldn't recommend it for you.

However, if you are a long time investor, Ethereum has quite a considerable roadmap when it comes to developing new projects.

Therefore, if you just want to buy some Ethereum, and hold it for long-term, it would be a good idea. You might choose to learn further on the Ethereum network, follow their news, and understand what kind of projects they are planning.

However, if you just want to invest in something that will hold your money, perhaps even increase it, once again, Ethereum can be an excellent choice.

**Ethereum**

Ethereum is currently the second biggest cryptocurrency, sits behind Bitcoin with a total market capitalization of 28 billion dollars.

| # | Name | Symbol | Market Cap | Price | Circulating Supply | Volume (24h) | % 1h | % 24h | % 7d |
|---|---|---|---|---|---|---|---|---|---|
| 1 | Bitcoin | BTC | $68,627,558,965 | $4,136.67 | 16,590,250 | $1,475,400,000 | 1.42% | 5.66% | 4.15% |
| 2 | Ethereum | ETH | $28,433,357,779 | $299.75 | 94,640,165 | $448,490,000 | 1.21% | 3.57% | 4.28% |
| 3 | Ripple | XRP | $7,856,308,107 | $0.204691 | 38,343,841,883 | $200,400,000 | 1.28% | 14.00% | 11.22% |
| 4 | Bitcoin Cash | BCH | $7,563,480,269 | $460.03 | 16,622,413 | $171,073,000 | 1.22% | 1.72% | 7.05% |
| 5 | Litecoin | LTC | $2,892,961,417 | $54.46 | 53,100,462 | $171,321,000 | 1.10% | 3.76% | 1.06% |

The current value of Ethereum is 299 dollars, however over a year ago was 13 dollars.

This means, if you have invested $100 a year ago, today it would worth around 2300 dollars. Its all-time high value was close to 400 dollars in June, then in September of 2017, however, I believe that Ethereum will rise once again.

## Ethereum Classic

Ethereum Classic is the 11th biggest cryptocurrency with the total market capitalization of 1.5 billion dollars.

The current value of Ethereum Classic is 12.62 dollars, however over a year ago was 0.61 cents. This means if you would have invested $100 a year ago, today it would worth around 2068 dollars. Its all-time high was close to 22 dollars in June and September similar to Ethereum.

As you see the return on your investment (ROI) are somewhat similar one to another, as well both Ethereum and Ethereum Classic had the all-time high value around the same time of the year.

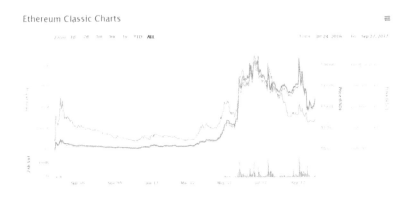

For further information about Ethereum and Ethereum Classic you may go ahead and visit their website on the following links:

Ehtereum
https://www.ethereum.org/

Ethereum Classic
https://ethereumclassic.github.io/

## Chapter 9 – NEO

Formerly known as Antshares, after a successful re-brand known as NEO. Also known as the Ethereum of China. Am I making this up? Well, it's NOE, themselves who publicly announced that they want to go ahead with a direct competition against Ethereum, and it's them who began spreading the word that they are the Ethereum of China.

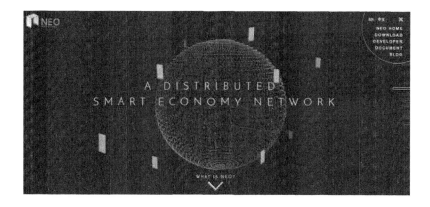

NEO, it's a distributed network, combined with smart contracts with digital assets and identities. First, NEO supports most advanced programming languages, such as JavaScript, NET, Kotlin, Phyton, while Ethereum only supports Solidity. This is already a massive advantage for NEO against Ethereum, and let me tell you why. One of the most significant problems in the whole crypto world is to find talented people who can deal with complex issues, and able to develop additional platforms or applications. When it comes to Ethereum developers, unfortunately there are fewer people who have knowledge of solidity, while there are lot's more people who know JavaScript, Phyton and other programming languages.

When it comes to an issue that requires fixing, there are more people available to participate with NEO. Simple as that. On the other hand, Vitalic recently was talking about this, and here is what his opinion about that.

He said that those traditional programming languages were designed without having blockchain in mind; however, solidity was strictly written for the blockchain. I am sure you can see already where all this goes!

There will be people who won't bother learning a new language, primarily if they have studied for years previously.

However, those who just begin to study, and have an option to learn a new language, they might choose solidity of course. Anyhow, I believe each programmer will have their own opinion about this, so let me move on.

NEO is China's first open source blockchain; therefore you can expect Chinese businesses to use the NEO blockchain in the future, to create many exciting technologies.

The platform itself focuses on building digital assets, in the way of converting traditional holdings to digital assets, then combining it with smart contracts. These digital assets will be decentralized by using the blockchain technology, and being protected by law, using digital certificates.

The primary purpose of NEO, is to map real-world assets, using smart contracts. This will enable individuals, as well businesses to manage those smart contracts efficiently, safely and legally. Additionally, NEO also focuses on decentralized commerce and digital identities.

Digital certificates will guarantee trusts, and the scalability issue, like Bitcoin has been facing. This is now resolved in the case of NEO, because NEO is using Lightweight stack VM (Virtual Machine).

NEO's algorithm is using Byzantine Fault Tolerance, also known as DBFT algorithm Instead of traditional proof of work or proof of stake, BDFT is using voting nodes for decentralization, which makes the system even more efficient. NEO is also Quantum Secured, by Lattice-based cryptography, meaning the blockchain will be stable against quantum computing attacks.

NEO coins also have another significant advantage, which is this: Once you will buy some NEO coins and you will hold it, you will continuously earn additional NEO coins.

This is somewhat similar to a proof of steak model, meaning, more NOE coins you have, more you will receive. NEO can support Dockers, which implies that NEO will be capable of lightening speed transactions.

The official NEO conference has been taken place on the 22nd of June, 2017, in Beijing.

At the time they also announced the official rebranding of Antshares to NEO, aka Antshares 2.0. In this conference, they have admitted without being shy that they will be the Chinese Ethereum, and their primary focus will be smart contracts.

Regards to the rebranding, NEO also changed their logo, as well their website to have a more professional style.

They used to have a logo that described a cartoon ant which have been looked funny, however with the new design; they will have a more serious look to concur even larger companies.

Under the old brand Antshares, they have managed to work with large companies such as Microsoft and Alibaba, however now with a more professional look, this marketing strategy will allow them to take the business to the next level.

The NEO team is already working with small startups to get the best out of the company, by doing multiple projects simultaneously. Some of the projects include:

• NEST Funds: This project is focused on NEO's blockchain to improve DAO attacks, which previously split Ethereum. This project, funds white hat hackers, whose primary job is to find vulnerabilities in the system, and keep on patching them, and making sure their system does not get hacked.

• Agrello: This project is focused on creating a smart contract for non-programmers. This means the NEO blockchain smart contracts are available to use for non-programmers too. Therefore, if you want to have a smart contract, and you have no programming knowledge, NEO can help you.

• Coindash: This project, is to offer advisory and prediction tools for new NEST investors. As I mentioned, NEST is to protect NEO'S assets, by continuously testing the platform for vulnerabilities.

• Binance Project: NEO works with the company called Binance. Binance is a digital asset exchange, that is currently competing against Poloniex. Poloniex is a cryptocurrency trading platform where no Fiat money can be traded. However, NEO will be listing some of their digital assets on Binance, and what you have to know is that Binance is indeed very selective when it comes to recording any cryptocurrency on their platform.

As you see, they are really focusing on securing new clients on their newly created NEO blockchain. Additionally, NEO has one of the top blockchain developer team that currently exist in China.

Now you know that NEO has a better potentials then Ethereum, and this is purely by able to take advantages on the whole Chinese market, you really should take a closer look. It's your choice, however, in my opinion, if Ethereum was capable of growing 30 times in value within a year, NEO might just be able to do the same.

Of course there is a possibility that NEO could exhibit even larger growth within the next year or even less then Ethereum.

NEO currently sits as a 9th biggest cryptocurrency, with a total market capitalization of 1.5 Billion dollars.

The value of NEO is currently 30 dollars, however, five months back, the value of NEO was around 0.2 dollars.

This means that if you were to invest in NEO in May only 100 dollars, today end of September 2017, your 100 dollars would worth 15000 dollars. As I mentioned earlier, if Ethereum was capable of reaching 300 dollars mark, NEO probably will do it too within a year or maybe even less.

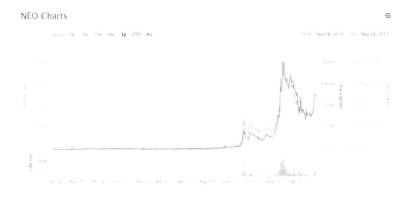

For further information about NEO, you may go ahead and visit their website at the following link:

https://neo.org/

## Chapter 10 – OmiseGo

OmiseGo, is another new coin that wants to disrupt existing financial institutions. Essentially what OmiseGo is an Ethereum based commercial technology, that has a peer-to-peer value exchange and payment service cross jurisdictions and organizational silos.

What this means, is people will be capable to send money across fiat and decentralized currencies. It is designed to enable financial inclusion and disrupt existing institutions.

The access will be made available to everyone, via the OmiseGo network and it's digital wallet framework. This project will only start in Quarter 4, 2017. It depends on when you read this book, however currently OmiseGo is an idea, nothing more.

Now, because it's through the Ethereum network, OmiseGo has been backed up by Vitalik Buterin. Not to mention that he is one of the top Advisor to the project, OmiseGo already has a massive hype on the market.

Because Vitalik is supporting this coin, everyone seems to go crazy about it, and you have admit, Vitalic who is the creator of Ethereum, is a genius, and probably would not provide his name to something that is baloney.

Currently, the financial industry and all the financial institutions could benefit from this project, by using the OmiseGo blockchain platform.

Through OmiseGo's network, which is connected to Ethereum, anyone will be able to conduct financial transactions. This will include all sort of payments like PayPal, deposits, Business to Business, Supply chain finance, Asset management, and much more, especially on demand services.

Any payments online, including banking, possible will be allowed to go through OmiseGo's blockchain platform. Furthermore,

OmiseGo can adapt to any transactions, meaning, in case the future of OmiseGo wants to become Fiat to cryptocurrency or vice versa, the platform will be able to adopt to it without any issue. Even though it's only an idea yet, still there are many financial investors already supporting and investing in OmiseGo's project.

The current volume for the coin already larger than usual, and this is making the value of OmiseGo stronger, literally every day. Of course, once the volume expanding, as well it's value getting stronger, it will continuously begin to create demand on both crypto and fiat market.

OmiseGo already has been launched in Thailand and Singapore, and continuously taking an interest by not only financial institutions, but the Thailand government itself. Then again, who is behind the company of OmiseGo? Well, let me elaborate on this topic now, so you will have a better understanding.

Omise, is already an existing company, and it's main purpose is to handle payments across Asia.

They are primarily specializing payment solutions for e-commerce businesses, by allowing to receive payments, from credit cards, or any traditional financial institutions. Omise is already operating in most of the Asian countries, such as, Vietnam, Thailand, Indonesia, Singapore, Malaysia, Philippines, and these are all small countries that have their own banking systems.

Because there are so many little businesses, there was always confusion with the payments, however, Omise stepped in, and this has made everyone's life lot easier by allowing fast transaction flows on the South Asian online markets.

They have started off in Thailand, then expanded to Japan, and now they are going to Indonesia, Singapore, and Malaysia.

OmiseGo is the next project for Omise, and what they are trying to do, is basically what they are doing best: combining banking services, and financial services within the blockchain.

They will create their own blockchain, and eventually merging decentralized exchange, with Liquidity provider mechanism, clearinghouse messaging network, and asset blockchain gateway.

All in one big project, not bad right? Cryptocurrencies brought forward the way to transfer value, between parties and even between nations, and OmiseGo will do all that now digitally in a system, where humans don't need to rely on a central body. OmiseGo is also trying to create an e-wallet solution, which would allow people in Southeast Asia to sign up.

Places like South East Asia, where people don't have bank accounts, or those who do not want to deal with the current financial system, they can sign up at OmiseGo, and they will be able to transfer value to other people on the OmiseGo network.

OmiseGo also accepts other cryptocurrencies, such as Bitcoin or Ethereum, and allows ways to transfer those currencies, and prove that people have those assets.

You might be thinking you don't need any of these services, however, what you have to understand, is in South East Asia there is a huge demand for services like this.

For example, in the Europian Union, most countries have the Euro currency, however, in South East Asia, they have all sorts of currencies; therefore it is complicated to transfer money between the countries.

Not to mention those hard-working people who live abroad, and trying to send money back to their home countries in South East Asia, the current swift system is charging very high fees; therefore OmiseGo will be an excellent solution for them.

However, you have to remember, that OmiseGo, is still a concept; therefore they have plenty of work to do to build, their own OmiseGo blockchian.

If you are ready to invest into OmiseGo, they are selling their tokens, which will provide a promise in their value in the future. The current token mechanism, called OMG token, is only representing a token; however, it will be used as a proof of stake in the network; therefore it will earn after each transaction, that will take place over time.

The current state of OmiseGo is they are trying to get approvals from all the Asian Governments to have a successful launch.

It is still plenty of work, and it will take some time for them to launch, however, because there is an existing successful company behind the currency, it is a token that you should take a closer look and consider investing.

OmiseGo is currently the 12th biggest cryptocurrency with the total market capitalization of 974 million dollars.

| # | Name | Symbol | Market Cap | Price | Circulating Supply | Volume (24h) | % 1h | % 24h | % 7d |
|---|------|--------|------------|-------|--------------------|--------------|------|-------|------|
| 1 | Bitcoin | BTC | $69,424,188,627 | | | | | | |
| 2 | Ethereum | ETH | $27,755,698,748 | | | | | | |
| 3 | Ripple | XRP | $7,487,440,248 | | | | | | |
| 4 | Bitcoin Cash | BCH | $7,309,486,882 | | | | | | |
| 5 | Litecoin | LTC | $2,824,279,675 | | | | | | |
| 6 | Dash | DASH | $2,501,770,166 | | | | | | |
| 7 | NEM | XEM | $2,143,847,000 | | | | | | |
| 8 | IOTA | MIOTA | $1,631,434,181 | | | | | | |
| 9 | Monero | XMR | $1,451,683,757 | | | | | | |
| 10 | NEO | NEO | $1,440,515,000 | | | | | | |
| 11 | Ethereum Classic | ETC | $1,219,870,144 | | | | | | |
| 12 | OmiseGO | OMG | $974,016,561 | | | | | | |

The value of OmiseGo is 9.91 dollars, however in July 2017, the value was around 0.45 cents.

This means if you would have invested in OmiseGo 100 dollars, in July, today: at the end of September 2017, your 100 dollars would worth about 2200 dollars.

I believe that OmiseGo has some real huge potentials in a very near future; therefore I would highly advise you to take an interest in their technology, and learn about them as much as you can.

Look out for their websites, and news letters, and try to stay up to date with their projects, as well achievements. In order to take a look at their website, you may go ahead and visit it by the following link:

**https://omg.omise.co/**

## Chapter 11 – TenX

The company called TenX, is about to create their own version of credit card; indeed it's the first of its kind in the crypto world, but let me dive in, so you can fully understand what TenX is up to.

At its very core, the TenX card allows to store cryptocurrency, but it let's you spend it in Fiat currency, therefore, you can have an immediate transfer at the point of sale.

It would work similarly to any other traditional bank card that allows you to pay; however you would be able to spend your cryptocurrency instead of your regular fiat currency. TenX also has an Android App; therefore you don't need the card, however, perhaps a plastic card has a smaller weight then your mobile phone. On the TenX website, there is a video of making payment in Dash, at the McDonalds, using TenX card.

https://www.tenx.tech/

The company already managed to make payments with Bitcoin, Ethereum, Litecoin, Dash, and promising much more cryptocurrencies to come. Their aim is to accept all cryptocurrencies eventually; however, it's still a long way to go.

One thing that currently makes me think about the project which is this. Each time when I go a payment, it is not in my best interests to pay with cryptocurrency, instead spend all my fiat currency and rather keep on holding to cryptocurrency.

However, they have an excellent solution for those who would panic like me, which is this: Each time you spend money, using the TenX system, you get cash back.

The advantages are for both the cardholders as well the token holders. If you are a token holder, meaning you only have TenX on your app, which is on your mobile device, you get 0.5% rewards monthly, calculated of all payment volume that is spent by users on the TenX card .

However, if you are a cardholder, your prize is 0.1% from each transaction people make with the TenX card. This is very motivational to many people, therefore, in my opinion, it will take off in the long run. Everyone like to make money while doing nothing right?

Because it's an income generator token, there will be much more people who will jump on board for sure.

There is also another way to make money with TenX, which is the capital gains that will come from the tokens, once people start to realize that those are an actual income generator tokens.

Additionally, there are some more great features to TenX, which for example an app that's working in conjunction with the card. Meaning, in case you lose your card, you would be able to block the card using your mobile app.

This is another great function. Again, in a traditional visa card system, if you would lose your card, you would have to call the bank to cancel or block your card, that could take you some time, of course, it depends on the time of your calling the bank. Anyhow, I never heard anything like it, so once again I am pleased with their technology.

**Passive income with TenX**

By 2018, TenX expects to start transacting, having at least one million users, and having 1 Billion dollars in transaction value.

By 2019, TenX expects to have 10 million users, and 20 billion dollars in transaction value.

By 2020, they are planning to have 50 million users, and 100 Billion dollars in transaction volume on their platform.

Now, this all sounds good, so let's do the math. Currently, the value of 1 TenX token is 2.67 dollars. Additionally, there are 105 million tokens in current circulation.

Also, let's imagine that you are investing $100 this year using the same rate. Of course, when you read this book, you might have to re-calculate it but, I am working with the numbers as of the 30th of September 2017.

If their predictions are right, by 2018, they will generate $2.5 billion in transaction fees, which means around 1.2 Billion dollars in payouts, this is if I assume 105 million tokens, with 100 dollars investments, it would equal to 1 dollar monthly passive income, which is around $12 yearly profit.

This is not a lot of money, considering I can sometimes make hundreds of dollars with Bitcoin when I am trading, however, let's move on.

Once again, if their numbers are correct, by 2019 they will have $20 billion in revenues, using the same figures, would primarily generate you 20 times more then year one, which is 240 dollars profit for year two.

It's hardy making any money for me but wait. So, for year three, it would also multiply by five times, of course, if their predictions are correct, meaning it would generate you a profit of 1200 dollars for year three.

Once again, it doesn't seem to be lot's of money, however, let's imagine that by three years time, the value of TenX will reach 50 dollars at very least. I personally believe that TenX value will go over 100 dollars within a year, however it's only an assumption, so let's stick with a 50 dollars mark within three years. This would mean that the value of all transactions also become 20 times more worth, saying if you would invest $100 this year while the cost of TenX is only $2.67, by 2020 you could potentially earn a passive income of 24000 dollars per year.

Once again, this is just my assumption, and with the considerations of investing only 100 dollars this year while the price is still low.

Anyhow, when you read this book, the price point will be different, which might be more or less, and your calculation will be completely different than it is today. However, the underlying math will not change.

To my mind, if I have to wait three years to make even only 10000 dollars of passive income a year, and I just have to invest 100 dollars today, it's a no-brainer.

Of course other cryptocurrencies can possibly generate even more over the years, however, $100 dollars to fund into something that unique and first of its kind, to me, I am fine with it. TenX cards are currently unavailable, due to the number of customers are reached the 10000 people.

What they said is that they prefer to look after their current customers first, making sure they are all experiencing an excellent customer service; therefore they can not allow additional customers for further notice.

Now, this is something great once again, of course if you are trying to jump on and they wouldn't let you do so, probably you will not be happy about it, but still, you have to give it to them: by looking after their existing customers is their priority, just makes me wonder how far they will reach in the near future.

**TenX Card**

In case you want to get your TenX card, first you have to download their free TenX mobile App, then once you have registered, you can ask for your TenX card.

Again, it depends on when you are reading this book, however currently they have issued 5000 Cards, already and if you want to have your own card, you have to be on the waiting list.

Once again, TenX could have registered more than 5000 people for a new card. But instead of failing to deliver to all, they choose to look after the first 5000 registered customers first.

TenX is currently the 22nd largest cryptocurrency, with a total market capitalization of 279 million dollars, and the current value is 2.67 dollars.

The trend of TenX is somewhat stable, however, because most of the plans are for the next few years, the real growth will be in the future in my opinion.

Even they have managed to have a working solution already, it is limited only to those who were first got involved and became part of their customers list.

However, once they announce that they will be able to accommodate another 5000 customers, the market capitalization can double, if not triple according to the current value.

If you look at the graphs, there is a possibility that once TenX had a high point, that was the time when those customers have been served, then because those people spend their money, there is a decrease in value.

In case you want to find out more on TenX, perhaps would like to sign up and order your TenX card, you may go ahead and visit their website at the following link.

https://www.tenx.tech/

## Chapter 12 – CIVIC

Civic is nothing but a decentralized digital identification platform. Sure, that doesn't make any sense, so let me dive in. What Civic does fundamentally, is secure and protect identities. Terms merely anywhere that you would require proving your identity Civic could help all those organizations.

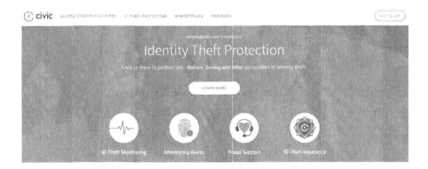

For example, when you go traveling and have to pass through the airport, it is very likely that you have to have your passport with you.

Another example when you have to clear something with your government, and they would require you to send your ID for verification purposes.

Or even any cryptocurrency trading platforms, where if you wish to register, they would need you to provide identifications. These methods also are known as: Know your Customer or in short KYC.

Each time when you must prove your identity, large companies costs lots of money, and human interventions are pretty much always necessary, and of course, these processes usually are taking a long time to complete. Civic is trying to fix all these issues, using blockchain technology.

What they are about to create is a decentralized network, where you as an individual, have your id and those companies or people who use Civic's blockchain technology will be able to ask you simple questions.

For example your full name, or date of birth, and necessary fundamental questions are all would be encrypted on the network.

So, all those questions that you have been asked and answered would be registered on your Civic's ID. This, of course, would be on the blockchain fully encrypted and secured.

Mainly, from now on, if you would go to the airport and you would provide your name and some other necessary information, they wouldn't need to use their own in-house build systems for background checks; instead, they could use Civic's blockchain. The reality is that not everyone understands yet how important that can be in the future, in term of saving time and money for everyone.

Primarily more you would use this technology, more you can become trustworthy on Civic's blockchain.

Now that's all nice, but does the Civic token worth anything? Basically, as the adoption continuous to the technology, the Civic token is used as a way of exchange, therefore as long as the adoption of Civic's blockchain increases, parallel the token's value will be increasing too.

Additional features Civic is participating are:

- Fingerprint technology
- Monitoring alerts
- ID theft monitoring
- Fraud support
- ID theft insurance

Civic has become the first company that turns identity theft protection into genuinely warning you before the identity theft hits you.

The current hacking that's going on worldwide is unbelievable, and 60 – 70 % of hacks are still focused on identity thefts, more specifically ID, passports, and social security numbers.

In 2014, 17.6 million US residents experienced identity theft; therefore Civic can become a huge industry, and of course many other companies will probably try to copy this idea in the future.

Civic is currently the 47th largest cryptocurrency, with the total market capitalization of 138 million dollars.

This amount of money itself is the proof there is a huge demand for such technology, therefore once again, I would highly recommend you to get into Civic.

The value of Civic is currently 0.40 cents, therefore there is still space for growth in the future.

If Civic makes you interested and thinks about to invest and participate in their blockchain network, I would highly advise you to take a look at their website, to do some further research.

**https://www.civic.com/**

## Chapter 13 – Guide to investing

Now that you have made it to this chapter, I would like to add some additional notes regards to investing. First of all, you might be reading this book at a later date, when you could find even better cryptocurrencies, which is fine.

However, when it comes to investing, the fundamentals that you must be aware will always be somewhat the same. My recommendation is that you should certainly look into all those cryptocurrencies I have introduced in this book; however, I am not an investment advisor, therefore do not blame me if you lose your money for any reason.

As I explained in some of my previous books, once you get involved in cryptocurrencies and begin to own some, you are your own bank. Therefore there is no one to blame but yourself, whatever happens.

In the world of Crypto, there is no guarantee for anything, and you must take responsibility for either you mistype an address and send coins to the wrong address, either your software wallet gets hacked because you fail to keep your cryptocurrencies on the hardware wallet. As always, I recommend to purchase the best and cheapest hardware wallet on the market called Ledger Nano S, which you can buy at the following link:

**https://www.ledgerwallet.com/r/e101**

Supply and demand move the value of each cryptocurrency. The market capitalization also has its part; however, the volume is what must be continuously in an increase, to have a better value of a specific currency.

Additionally, I have realized that all cryptocurrencies are immune for news, such as any war, or fiat currency issues or even natural disaster. However most of the coins are still highly volatile, and it is because not everyone using cryptocurrencies yet.

Once there will be more adoption around the world, less volatile the crypto market will become.

Buying in a coin quickly, without consideration, is always wrong. Investing in something that is not 100% clear to you, please do yourself a favour, and do not proceed.

Technical analysis and full awareness of a specific coin are always critical. Additionally you must always look into the future of the companies plans before investing.

Let's take an example, what's important before you would invest in any cryptocurrency. Most of these questions that you must ask are also relevant to other business models, not only cryptocurrencies. So let's begin.

## WHAT YOU MUST KNOW BEFORE INVESTING

1.
When taking an interest in a specific cryptocurrency, first, you must understand what it is: is it a currency, application, or a platform. Download their whitepaper regards to the project directly from their website, read it and try to understand it!

2.
Who is behind the project? If you see well-known names, that's already a great news; however, it doesn't mean that the cryptocurrency will work out well; yet, you should check the names as soon as you can.

If there are known names, double check that those people are involved in that project. There are many scam ICO-s where known people were added to the project contact list as senior advisors, then it turned out that was all fake. Do not get scammed, do as much research as possible. Having looked at one or two resources sometimes aren't enough.

Additionally, even there are known people backing the project, and it's true, still do your research and ask all the questions on the rest of this list below.

There have been projects when for example Bitcoin Jesus was known as Bitcoin Anti Christ. Also seen other projects, were recognized and respected people have only created hype around a new coin, so they can play their pump and dump game, so they can increase their own finance.

I am not pointing fingers, but warning you: instead of following someone idea, you must make your own decision.

If you see a well-known person, who takes a keen interest in a particular project in the way of advertising it, research and try to understand why he or she does it.

Why is it so beneficial? - is it for friendship and only helping marketing for a friend? is it for the project so they really want to help people? , or just marketing for a purpose of potential financial gain? Financial gain can be achieved when a new coin is about to born by a forced hardfork on an actual successful currency.

3.
Next, you must understand, if it's only an idea such as an ICO, or it might have passed an ICO already. If, it has given an ICO, was it a successful quick ICO, or it was slow and seems like a pump and dump only.

4.
What does the cryptocurrency do? What is their goal? You must understand it entirely, and you can test it by explaining it to your friend for example, who should confirm that he or she understands what you are talking about.

5.
Storage of the currency! Do they have a cryptocurrency wallet, and if the answer is yes, how easy is to use that wallet? If they don't have one yet, find out what's the plan for creating one. If they are not clearly describing their plan for their own wallet, and future storage, or should move on.

6.
What sort of marketplace they are targeting? For example, if it's currency, who is it for, and why is it different then any other existing

currency that already on the market? If it's application, who is it for to use that specific application? If it's a platform, what kind of apps is that platform for, and who would use those applications?

For example the platform for only one specific application? – or many application in the same niche, or multiple niches?

More specifically, for example: is it for all sort of gaming platforms only? – or is for all gaming platforms as well all on-demands music, and video applications too! In case it's application or platform, is it scalable, or for limited users only?

Once again, you want to see dynamic planning, as you must be aware that the market will move on, therefore they must have planning for scalability, and not only for one specific solution.

7.
It all depends on option 5, however, whatever is the plan for the cryptocurrency, you have to think about, and ask the question: if is it a realistic plan?, Is it achievable?

In case you like the idea, but you are not sure if is feasible, and sensible plan, you must seek advice. You must ask someone who is experienced in that specific field and has an in-debt knowledge on the topic or topics. Like I mentioned in option 6: they should have plan for scalability sure, but if it's unreasonable or seem like an overpromise, you must seek professional advice.

8.
In case you can not find someone who has an in-depth knowledge on a specific topic, then you have to investigate the team behind the cryptocurrency. Generally on their website, when you click on about, or contacts, you should have all the names behind the project. Your job, is to Google their names and see if they have relevant experience on the field.

This is another turning point, as if you can not find enough information about the team, neither a specialist who can confirm you that the project has realistic plans, you should move on, and analyse another cryptocurrency.

There is no point to think that is a good idea, and what they plan is incredible, especially if there is no proof of someone has relevant experience on the field.

Do not get emotional, because you like the idea, anyone can have a great idea, it is not the idea, but the action that will take the concept to success. If there is no one, or seem to be unqualified people behind the project, you must move on. Do not forget that is a business investment, not a donation. Sorry, but once again, emotions should not be used when you think of an investment.

In the other hand, if the team seem to be an expert in their field, having years of experience, and relevant background projects, you might just going to hit the jackpot.

But, how do you know if it's the right team behind the project? Well, you have to ask them. Typically as I mentioned when you click on the About Us, or Contacts menu on their website, you should be able to see their names, and probably there is a way to contact them too.

9.
Contact the team. You probably think that's an overdue, because lot's of people have became millioners without talking to anyone, which is true. But please do not forget that in the same time at least 10 times more people lost their money in the same time.

You hear success stories all the time, but people who make mistakes, and lose money, they don't go around telling people how much they lost, instead they tell no one. Additionally, not all but most people who going around telling everyone their success stories, what they don't tell you is that how much money they have lost before their success.

You have to understand that before success, there is a lot of trial and error, and if you still reading this book, you really should be paranoid and know exactly who or what company you are about to invest your hard earn money. So, back to contacting the team!

Once you see that there is a way to get in contact with the team, you should ask the following questions:

- What kind of licenses do they need, and how long does it take to get them to start the project?
- How do they deal with the regulations within that specific country they are launching their products?
- How many people in the team, and who is going to be responsible with what projects, and how they are allocating their time for those, as well how often they are discussing, and dealing with issues?
- What are the worse issues they must deal with, and how do they plan to concur them?

- If there are issues within the project, how much sit back or delay is expected in terms of product launch?
- How they are going to deal with the marketing and advertising while there is a delay? Are they doing themselves or hiring professionals?

You might create your own e-mail template, and use the same one, and send it to multiple companies, and see who replies and how. Things that you have to look into is this:

- How fast they are replying?
- Can they answer to all your questions? – or they only answer some of them
- Are the answers are accurate? –or just wishy washy bla-bla content!
- Do they welcome you to take interest in their projects? –or it seems like they don't want you to write them again!

10.
Information gathering. Once you contact the team members, perhaps the CEO, do you get a reply? If yes, how fast and accurate is the response? Does it answer all your question in a trustworthy way? Many companies spend lots of marketing, which can be great, as this would

most probably tell you they are not afraid to talk about their plans or existing success, and failure too. One of the best news channels, in my opinion, is YouTube, where often team members of the project or even the CEOs talk about what they are currently working on.

These people should sound like having the vast technical knowledge, proud of their work, and what they are trying to achieve.

Twitter, Facebook is also lovely to have, especially if it's frequently updated with news of their tasks and communication to the supporting community. Some great projects even have an organized blog, where most times you get answered to all your questions.

When it comes to a problem, and you see that the CEO is brave enough to go public and talk about it, typically it's a great sign. This would indicate two things.

One is that they are not just scamming people telling everyone that everything is roses, like a sales pitch. The other is that you could probably hear the honestly, when they explain what challenges they have, and how they will approach concerning fixing it permanently.

11.
Count the followers. Check their social media platforms, such as Facebook, Twitter, and YouTube, where you can see how many people likes, subscribes or follows their sites.

Don't forget this: supply and demand. If there is no interest in the first place, there will be no investors. If there are no investors, the value will stay low.

Not only the value will remain low, but if the volume decreases, the chart will look bad, meaning even those who considered it before, will probably forget it and move on to another projects. Another significant issue here if there are no followers, and not enough investors, the project might not going to get the right amount of money to start the project.

I have talked about this before, so here it is again. Even though the plan is excellent and realistic; an the team members are blockchain and other project related field experts, if there is not enough money to support the project, no matter what, it will fail!

And if you have invested some money into a failed project, well, there are no refunds. Once again, you want to see thousands, if possible 10s or even 100s of thousands of followers. More is better of course. Eventually, all those followers will become investors, or at least a large percentage of them.

I have looked at some Twitter statistics for few of the most popular coins for your reference, and when you look at the followers, somewhat the volume, the market capitalization and the value of these cryptocurrencies are reflecting on their followers.

Bitcoin
343K Twitter followers
$71,635,216,831 Market capitalization

Ethereum
165K Twitter followers
$28,462,144,479 Market capitalization

Litecoin
114K Twitter followers
$2,882,281,638 Market capitalization

Monero
71K Twitter followers
$1,405,046,001 Market capitalization

Zcash
32.6K Twitter followers
$626,670,365 Market capitalization

DASH
80.6K Twitter followers
$2,379,113,226 Market capitalization

PIVX
24.2K Twitter followers
$197,454,569 Market capitalization

NEO
99.8K Twitter followers
$1,795,505,000 Market capitalization

OmiseGo
69.2K Twitter followers
$810,756,650 Market capitalization

TenX
47.4K Twitter followers
$214,701,165 Market capitalization

Civic
36.4K Twitter followers
$108,842,567 Market capitalization

Filecoin
12.1K Twitter followers

No Market capitalization yet >>> more on this later

12. Roadmap. Most companies should have a roadmap on their website, which represents the plan for future development. The roadmap should include dates not only for the future, but previously achieved projects too, predicted achievements, and comments for delays, challenges and so on.

If there is no roadmap, or the roadmap doesn't make sense, for example, the plans are unrealistic; you should probably better if not investing.

However, if the roadmap includes all projects that are achievable, and it has been written transparently so the investors can understand it, that's great news, but you must follow it, making sure that you are aware of the technology that you have invested.

Some occasions, they might list some issues that they have to figure out how to overcome, and these times you could even see that the value of that cryptocurrency is decreasing.

If you believe they can overcome this issue, it might be a good time to invest, while the value is down to their coins. In the other hand when they announce an achievement, you should also get ready to see an increase in volume as well in value.

13.
Who is the competition? You must watch out, and understand exactly what other similar cryptocurrencies are already on the market, and once again, you should think about if is a realistic project to compete with those.

For example, if there is a new currency that has a sole purpose of payments, I wouldn't invest, and here is why:

There are some great cryptocurrencies that already serve well its entity, such as Bitcoin, Zcash, Dash, Monero, Litecoin.

To come up with another new currency that would fix all the existing issues, still, wouldn't have much of change against these big players.

## Overview

I have done plenty of research, and hope my advice will serve you well in your future when it comes to investing in cryptocurrencies.

However, I would like to point out few things. First of all, no matter what will happen with the coins that I have talked about, you must always be extra vigilant, and make sure that you **do not get scammed**.

Also, you do whatever you wish; therefore you might want to invest into other cryptocurrencies, that I have not mentioned in this book, it's okay, it's entirely up to you.

There should be no one to tell you where to invest your money.

I am not a financial advisor; however I did my best to provide you the best technical analysis for one of the best blockchain based cryptocurrencies, and I hope that you will be successful on the Crypto market.

Lastly, I would like to point out some more information that you should be aware at all times!

1st
If you do not do your research, there is a high probability that you will lose your money. Therefore, study, research, and research before taking any action.

2nd
Understand the potentials for both the risk involved, as well the possible ROI (Return on investment)

3rd
ICO-s for experienced traders and investors, those, who have done multiple technical analysis, therefore you might want to start to invest into cryptocurrencies that are already existing on trading platforms.

In case you want to play safely, like I do, do not go crazy trading all over the place, instead invest in existing popular cryptocurrencies, like most of them I have explained in this book. Either way, you must make your own decision, and take responsibility for all your actions.

## Bonus Chapter – Filecoin

In my journey of researching, I have came a cross with lot's of scam coins, and some proper rubbish time waste, but this has also provided me excellent opportunity to learn how to recognize something unique.

Again, when I say unique, I am not talking about an idea itself, but an overall feel of the team behind the proposed project, and those list of considerations that I have explained in the previous chapter.

Yes I did come across with Onecoin, and Jesuscoin too, but please, stop the madness! Ok, in a case of Onecoin was really a tricky one, simple because the marketing was well organized, they were able to scam lot's of peoples and steal their money.

Anyhow, those that participated in marketing it as well the creators have been sent to jail already. But Jesuscoin?! I am out of words that people fall for this, it's just crazy! Apperently they have raised more then 200 Ether which is equals to 60K dollars. Anyhow, I will refuse to waste any further time on scams, instead let me introduce something that really made me to think: Filecoin. Filecoin is an idea for the decentralized storage network. Well, the concepts might be simple; however, the idea is a genius. But really, what else is there? Well, everything! OK, fine not everything, but when you think of the internet, you can use it for anything that's related to data right?

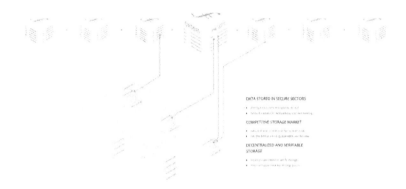

In the last 25 years, we have used the internet as our storage, as well as a point of medium to share everything with each other. This would include images, videos, or any kind data files.

We would use websites for these purposes, such as Facebook, Twitter, or YouTube, or even our websites and blogs so that we could share our experiences, or knowledge with the rest of the world.

The reality is that Facebook or YouTube do not belong to you. For example, if any of these websites would request to pay a specific fee, in order to continue participating, what could you do? The way the internet was built is that, there is a single point of failure everywhere, with every company there is.

So if for example Facebook would get hacked one day, all your data that you have ever shared could be lost. I mean if the banks get hacked that's where solutions like Bitcoin can resolve, however, when we are talking about our pictures, videos, or messages, you realize that we are talking about a large data that is currently out there and centralized.

One of the biggest cloud service providers are, of course, Amazon Web Services and Microsoft. Now, these are huge companies and specializing in accommodating other large organizations, so they can store their data in the cloud. Some of these data can be such as:

- Network Infrastructure,
- Server Infrastructure
- Back up infrastructure,
- E-mail infrastructure
- Payroll Infrastructure
- HR employee data infrastructure
- Project management infrastructure and much more.

There is a considerable amount of data that required backing up in every single minute of the day. This is of course already accommodated by many companies; however large organizations have been cautious of having only one back up, using just one company.

What most businesses do, is to rent multiple data centres in different locations to make sure their data is more secured, in case one of the companies they use get compromised.

As you see the current solutions are costly, taking a long time to get it right, and with too much security, it is challenging to access data.

This is where Filecoin comes into the picture. They believe that is incredibly important to improve the properties of the internet, with the idea of the technologization of the peer-to-peer network.

The genius behind Filecoin is Juan Benet, who has previously run another project called IPFS. IPFS is a protocol that already removed the central point of failure, to improve the resilience of the internet, and to provide information in a stronger sense or permanence.

IPFS has an amazing community of developers that are interested in building decentralized technologies.

The first phase of the work of decentralizing the web was to build IPFS. The next stage, improving on the foundation is a new protocol, called Filecoin. With the rise of protocols like Bitcoin and Ethereum, they have seen massive decentralized networks formed, where anyone can dedicate their computer power to run the network, and earn crypto tokens by doing so.

Mining for tokens is so profitable that these networks have spread globally and created the worlds most powerful computing systems. They have realized they can use the same kind of structures of Bitcoin to create a decentralized storage network.

There is a massive amount of available maintenance storage out there, and their vision is to create a decentralized storage market, a system, where all the convenient storage can be sold on the market.

With this vast amount of supply, join the network that is not currently connected, they can optimise their resources and drop the cost of storage. There is a native token to the system, called Filecoin. Filecoin is the medium of exchange for the decentralized market. The way it will work is that users would spend Filecoin to hire the network, to store their files, then storage providers get paid in Filecoin, which would scale, by how much storage they would add, and how close they are to their users.

One of the powerful thinks about Filecoin, is that it allows individuals to be incentivized to act as parts of the network.

Each node would contribute their part, for which they would get paid. All those who participate, would create a network that is more powerful than the systems that we have today.

Filecoin can be exchanged to Bitcoin, Ethereum, Dollars, Euros and so on. However, users can hold on to their Filecoin, which may appropriate as the network grows.

Anyone who uses the technology can become an investor in the technology. Filecoin is going to re-imagine the way, data stored for delivery.

Same as with Bitcoin that can be mined anywhere in the world, people will realize that they will be able to mine Filecoin too, by having good hardware, which not only going to be profitable but will be able to help their local community. The Filecoin ICO has happened already, and not only were able to raise the amount that they were required, but they did it in less than 30 minutes.

This is record-breaking ICO that ever happened, so I am sure that you can see it's a very serious business model.

Additionally, if you wish to participate, it's not as easy as you might think. So far, all the cryptocurrencies I have talked about are possible to purchase through various cryptocurrency trading platforms; however Filecoin is not live yet, therefore it is very difficult to become an early investor.

Basically, you have to be an accredited investor to participate, however, once they go public; hopefully anyone can invest.

If you interested to read the full report, go ahead and visit the link at coindesk website:

**https://www.coindesk.com/257-million-filecoin-breaks-time-record-ico-funding/**

If you are an accredited investor, you can apply to invest in Filecoin at the following link:
**https://coinlist.co/**

Additionally, to reach their website you can follow the link provided:
**https://filecoin.io/**

Be aware that Filecoin is still not listed on coinmarketcap website, however because the ICO already has ended, usually, can take at least one month to finalize the token sale and arrange with the exchanges.

So, most probably by October 2017, you will be able to see Filecoin on few exchanges; however I truly believe that the value will increase very quickly, therefore earlier you can invest better deal you will make.

I am not looking at Filecoin as a rulet table, instead as a long term investment.

# Conclusion

Thank you for purchasing this book. I hope the content has provided some insights into what is really behind the curtains when it comes to the future of money.

I have tried to favour every reader by avoiding technical terms on how to invest in Cryptocurrencies.

However, as I mentioned few times, to fully understand how Bitcoin and most Cryptocurrencies work, you may choose to read some of my other books on Blockchain as well on Bitcoin.

*Blockchain – Beginners Guide - Volume 1*

*Blockchain – Advanced Guide - Volume 2*

*Bitcoin Blueprint - Volume 1*

*Bitcoin Invest in Digital Gold- Volume 2*

*Cryptocurrency Trading – Volume 1*

The Blockchain books are focusing on the underlying technology of Bitcoin.

Blockchain Volume 1, is a beginners guide that teaches you to have some basic understanding of the technology; however, Volume 2 is very technical.

Still, I did my best to use everyday English and making sure that everyone can understand each of the technologies and their importance on the Blockchain.

Bitcoin blueprint focuses on Bitcoin basics; however, I was able to add some interesting and more advanced topics around the future of payments, using reputation systems with Bitcoin-Blockchain technology.

The Cryptocurrency Trading book is focusing on market terms, that you must be familiar with to stay on top of the quickly manipulated market.

However, the primary focus of this book is to provide techniques and strategies when it comes to cryptocurrency trading. Additionally, it focuses on building the most profitable cryptocurrency portfolio.

Some of my upcoming books on Bitcoin will provide more details on Bitcoin mining, profitability, and what precisely miners do behind the scene. Furthermore, how miners are often manipulating the market, what techniques they use, and how they try to control mining pools, using super expensive ASIC mining hardware.

I will provide information and secrets on Chinese Bitcoin miners, as well their exclusively produced ASIC mining rigs, that no other miners capable of using.

Lastly, if you enjoyed the book, please take some time to share your thoughts by post a review. It would be highly appreciated!

Printed by BoD in Norderstedt, Germany